SCIENTIFIC AMERICAN™

Critical Anthologies on Environment and Climate™

CRITICAL PERSPECTIVES ON
WORLD CLIMATE

edited by Katy Human

The Rosen Publishing Group, Inc., New York

Published in 2007 by The Rosen Publishing Group, Inc.
29 East 21st Street, New York, NY 10010

First Edition

Library of Congress Cataloging-in-Publication Data

Critical perspectives on world climate/edited by Katy Human.—1st ed.
    p. cm.—(Scientific American critical anthologies on environment and
climate)
Includes bibliographical references and index.
ISBN 1-4042-0688-4 (library binding)
1. Climatic changes. I. Human, Katy. II. Title. III. Series.

QC981.8.C5C775 2007
551.6—dc22

                                                                2005033235

Manufactured in the United States of America

On the cover: A fishing boat navigates past a receding glacier in the
Ilulissat fjord on Greenland's western coast in September 2004.

CONTENTS

Introduction   5

1. The (Pre)History of Climate Change   11
"Abrupt Climate Change" by Richard B. Alley   12
"Core Questions" by Marguerite Holloway   31
"How Did Humans First Alter Global Climate?"
    by William F. Ruddiman   35
"Snowball Earth" by Paul F. Hoffman
    and Daniel P. Schrag   50

2. The Role of Vegetation and Methane
   in Climate Change   73
"Deserting the Sahara" by Sarah Simpson   74
"Methane Fever" by Sarah Simpson   78
"When Methane Made Climate" by James F. Kasting   82
"In the Heat of the Night" by Tim Beardsley   100

3. The Present and Near-Future Effects of
   Global Warming   104
"Stormy Weather" by Mark Alpert   105
"Meltdown in the North" by Matthew Sturm,
    Donald K. Perovich, and Mark C. Serreze   109
"Is Global Warming Harmful to Health?"
    by Paul R. Epstein   122
"The Human Impact on Climate" by Thomas R. Karl
    and Kevin E. Trenberth   144

4. The Ongoing Debate over Climate Change   157
 "Behind the Hockey Stick" by David Appell 158
 "No Global Warming?" by Kristin Leutwyler 165
 "Hot Words" by David Appell 169

Web Sites 174
For Further Reading 175
Bibliography 177
Index 179

Introduction

During the summer of 2005, temperatures soared above 100 degrees Fahrenheit (38 degrees Celsius) in Denver, Colorado, for five days in a row, breaking records each day. Power companies struggled to meet the demand for electricity created by air conditioners and fans that were running constantly during the heat wave. Newspapers ran stories about the worst jobs to have to perform in the heat: roofing, laying asphalt, and landscaping. One of the best jobs was said to be working in the National Ice Core Laboratory, about fifteen miles (twenty-four kilometers) from downtown Denver, where temperatures are a constant −30°F (−34°C). On the streets, on front porches, and in air-conditioned restaurants, everybody was talking about the weather. One often heard the term "global warming" muttered, drawing weary nods each time.

There is no way to attribute any single weather event to human-caused climate change, however. A hot spell may occur randomly or be triggered by any number of other weather events around the world. Climate change is, by definition, a decades-long process. Researchers studying shifts in the climate look at trends that have occurred over the last century, the last thousand years, the last 100,000, and the last million.

However, in recent times, as temperatures across the planet have been breaking heat records year after year, scientists and nonscientists are beginning to pay attention. Many people now argue that climate change is the phenomenon that will most dramatically alter the lives of people on this planet in coming decades. As the articles in this anthology make clear, climate change will affect where people can live comfortably and safely. It may trigger more—and more severe—natural disasters such as floods and storms. It will alter agricultural patterns and opportunities, shifting the places where we can grow corn or rice or collect maple syrup and changing the length of growing seasons. Climate change may even affect public health, expanding or otherwise altering the range of disease-carrying organisms, or setting up fertile conditions for the spread of contagious diseases.

In these articles, you will be introduced to the general and emerging scientific consensus regarding climate change: It's happening, humans are causing much of it, and it is likely to eventually melt much of the planet's ice, with widespread and potentially devastating repercussions. You will also gain insight into the finer details of climate research and the physical processes and forces driving climate change—the limits to what scientists know, how they do their research, and some of the feedback cycles that contribute to the changes already happening on the ground.

Before you read the articles, it is worth reviewing a brief summary of the basic mechanics of greenhouse warming and some of the feedback cycles involved in climate.

On a very warm day, the interior of a parked car with its windows rolled up gets brutally hot, much hotter than the outside air. Incoming solar radiation warms the interior surfaces of the car, which heat up and reradiate energy, but in a new form. The radiated energy can not escape through the car's closed windows. This trapped radiant energy is like greenhouse gases in the atmosphere. Greenhouse gases, from carbon dioxide and methane to water vapor, let in incoming radiation, but they trap reradiated heat near the earth's surface, preventing much of it from escaping back to space. With no greenhouse gases in the atmosphere, the earth would be a brutally cold place, a "snowball planet." With too many greenhouse gases in the atmosphere, however, heat can build up on the planet's surface, triggering other sorts of changes, which appear to be equally devastating for life.

Negative feedback cycles (when a change in the earth-atmosphere system results in a chain of events that eventually lessens the impact of the initial change) can help keep a system, like the earth's climate, from getting too far out of balance. Imagine that increased warming puts more water vapor into the air, which makes the planet cloudier. Scientists once hoped that this would start a negative feedback cycle: Clouds would shade the planet from the sun, reducing some of the incoming solar radiation that was triggering the warming in the first place. As a result, things might cool down a bit, water vapor could drop back out of the atmosphere, and a short cycle of warming might start again, only to be cut off by the formation of more clouds.

Earth's atmosphere, of course, is a far more complicated system than this model allows for, with far more variables and forces at work. For example, some types of clouds shield sunlight well, while others let it pass right through to the surface, trapping heat down low. And water vapor itself is a greenhouse gas, so a moister atmosphere can actually trigger positive feedback cycles—the more dangerous sort from a global warming perspective. A positive feedback cycle is one in which a change in the earth-atmosphere system results in a chain of events that increases the impact of the initial event.

The classic positive feedback cycle in climate science involves the ice at the planet's poles. Ice and snow are white, and they reflect incoming solar radiation, bouncing it straight back to space. By contrast, open water or land appear dark from space. They absorb incoming solar radiation and heat up in response to it. So if warming begins to melt ice, more radiation-reflecting surfaces will turn into radiation-absorbing open water or dark land, allowing still more warming to occur.

You will see many examples of both positive and negative feedback cycles in the articles that follow. At this point, these various models of climate change remain somewhat speculative and theoretical. New data is being gathered every day in an attempt to comprehend the processes under way, but the processes themselves remain poorly understood and unpredictable. The science of climate change itself changes rapidly, partly because of advances in computer power that allow for more sophisticated and accurate climate modeling, and partly because conditions on the earth are changing quite rapidly. During the summer of 2004, newspapers ran story after story about melting ice across the planet, from the poles to midcontinent glaciers and snow-fields. Scientists predicted that the glaciers found in Glacier National Park would disappear by 2050 or earlier. This process was already well under way, as they watched Alaskan and Antarctic glaciers surge into the ocean as the ice dams once holding them in place steadily disintegrated.

In that same year, the debate about the link between global warming and hurricanes intensified, after four hurricanes (Charley, Frances, Ivan, and Jeanne) hit the Atlantic and Gulf coasts of the United States, causing more than $45 billion in damages. The debate reignited and crescendoed in late August 2005, when one of the most destructive Atlantic storms in history, Hurricane Katrina, made landfall just east of New Orleans, Louisiana, devastating that city and other communities along much of the Gulf Coast. Just three weeks after Katrina wreaked its havoc, Hurricane Rita, one of the three strongest Atlantic hurricanes in recorded history up to that point, struck farther west, pounding the Texas and Louisiana coasts and prompting the largest evacuation in Texas history.

You are likely to continue reading news about climate change in the months and years ahead. Melting ice will likely continue to be a big story, as will rising sea levels and increasingly destructive weather. The articles in this anthology provide an invaluable and in-depth grounding in the contentious issues and complex science discussed in the breaking environmental and climatological news so prevalent these days. They do a superb job of stating where we are right now and where we seem to be going, while often appropriately raising the alarm and calling for pragmatic and insightful preparations for an almost certain future of significantly altered climate. How the information contained in these and similar studies is used by humans may very well determine the fate of the race beyond the twenty-first century. *—KH*

1 | The (Pre)History of Climate Change

Richard B. Alley is a great example of a scientist who, in addition to conducting top-notch research, can also communicate the abstractions of theory and data in concrete, readily understandable terms. For example, in this article, he compares climate-altering human activities to the destabilizing motions a person can make in a canoe. With our continuous emissions of greenhouse gases, we are rocking a boat that is already unstable, Alley writes. Ice-core records reveal that abrupt changes in the climate have occurred in the past, such as a sudden chill that may persist for millennia, or an intense and unexpected warming.

If we persist in activities and practices that generate greenhouse emissions, we are likely to push the boat to a tipping point, Alley believes. But while a canoe rider can sense when he or she is leaning too far over the side and can then move back toward the center of the boat to steady it, people have not been able to learn enough about climate change to identify all the invisible tipping points in the climate system. We will not know when we have gone too far, past the point at

*which we could have corrected the balance. By
the time we reach that point, it will be too late to
prevent capsizing. —KH*

"Abrupt Climate Change"
by Richard B. Alley
Scientific American, November 2004

In the Hollywood disaster thriller *The Day after
Tomorrow*, a climate catastrophe of ice age proportions
catches the world unprepared. Millions of North
Americans flee to sunny Mexico as wolves stalk the
last few people huddled in freeze-dried New York
City. Tornadoes ravage California. Giant hailstones
pound Tokyo.

Are overwhelmingly abrupt climate changes likely
to happen anytime soon, or did Fox Studios exaggerate
wildly? The answer to both questions appears to be
yes. Most climate experts agree that we need not fear a
full-fledged ice age in the coming decades. But sudden,
dramatic climate changes have struck many times in
the past, and they could happen again. In fact, they are
probably inevitable.

Inevitable, too, are the potential challenges to
humanity. Unexpected warm spells may make certain
regions more hospitable, but they could magnify
sweltering conditions elsewhere. Cold snaps could
make winters numbingly harsh and clog key navigation
routes with ice. Severe droughts could render once
fertile land agriculturally barren. These consequences

would be particularly tough to bear because climate changes that occur suddenly often persist for centuries or even thousands of years. Indeed, the collapses of some ancient societies—once attributed to social, economic and political forces—are now being blamed largely on rapid shifts in climate.

The specter of abrupt climate change has attracted serious scientific investigation for more than a decade, but it has only recently captured the interest of filmmakers, economists and policymakers. Along with more attention comes increasing confusion about what triggers such change and what the outcomes will be. Casual observers might suppose that quick switches would dwarf any effects of human-induced global warming, which has been occurring gradually. But new evidence indicates that global warming should be more of a worry than ever: it could actually be pushing the earth's climate faster toward sudden shifts.

Jumping Back and Forth

Scientists might never have fully appreciated the climate's ability to lurch into a radically different state if not for ice cores extracted from Greenland's massive ice sheet in the early 1990s. These colossal rods of ice—some three kilometers long—entomb a remarkably clear set of climate records spanning the past 110,000 years. Investigators can distinguish annual layers in the ice cores and date them using a variety of methods; the composition of the ice itself reveals the temperature at which it formed.

Such work has revealed a long history of wild fluctuations in climate-long deep freezes alternating with brief warm spells. Central Greenland experienced cold snaps as extreme as six degrees Celsius in just a few years. On the other hand it achieved roughly half of the heating sustained since the peak of the last ice age—more than 10 degrees C—in a mere decade. That jump, which occurred about 11,500 years ago, is the equivalent of Minneapolis or Moscow acquiring the relatively sultry conditions of Atlanta or Madrid.

Not only did the ice cores reveal what happened in Greenland, but they also hinted at the situation in the rest of the world. Researchers had hypothesized that the 10-degree warming in the north was part of a warming episode across a broad swath of the Northern Hemisphere and that this episode enhanced precipitation in that region and far beyond. In Greenland itself, the thickness of the annual ice layers showed that, indeed, snowfall had doubled in a single year. Analyses of old air bubbles caught in the ice corroborated the prediction of increased wetness in other areas. In particular, measurements of methane in the bubbles indicated that this swamp gas was entering the atmosphere 50 percent faster during the intense warming than it had previously. The methane most likely entered the atmosphere as wetlands flooded in the tropics and thawed up north.

The cores also contained evidence that helped scientists fill in other details about the environment at various times. Ice layers that trapped dust from Asia indicated the source of prevailing winds, for instance.

Investigators concluded that the winds must have been calmer during warm times because less windblown sea salt and ash from faraway volcanoes accumulated in the ice. And the list of clues goes on [see "Greenland Ice Cores: Frozen in Time," by Richard B. Alley and Michael L. Bender; SCIENTIFIC AMERICAN, February 1998].

Intense, abrupt warming episodes appeared more than 20 times in the Greenland ice records. Within several hundreds or thousands of years after the start of a typical warm period, the climate reverted to slow cooling followed by quick cooling over as short a time as a century. Then the pattern began again with another warming that might take only a few years. During the most extreme cold conditions, icebergs strayed as far south as the coast of Portugal. Lesser challenges probably drove the Vikings out of Greenland during the most recent cold snap, called the Little Ice Age, which started around AD 1400 and lasted 500 years.

Sharp warm-ups and cool-downs in the north unfolded differently elsewhere in the world, even though they may have shared a common cause. Cold, wet times in Greenland correlate with particularly cold, dry, windy conditions in Europe and North America; they also coincide with anomalously warm weather in the South Atlantic and Antarctica. Investigators pieced together these regional histories from additional clues they found in the ice of high mountain glaciers, the thickness of tree rings, and the types of pollen and shells preserved in ancient mud at the bottoms of lakes and oceans, among other sources.

The evidence also revealed that abrupt shifts in rainfall have offered up challenges rivaling those produced by temperature swings. Cold times in the north typically brought drought to Saharan Africa and India. About 5,000 years ago a sudden drying converted the Sahara from a green landscape dotted with lakes to the scorching, sandy desert it is today. Two centuries of dryness about 1,100 years ago apparently contributed to the end of classic Mayan civilization in Mexico and elsewhere in Central America. In modern times, the El Niño phenomenon and other anomalies in the North Pacific occasionally have steered weather patterns far enough to trigger surprise droughts, such as the one responsible for the U.S. dust bowl of the 1930s.

Past as Prologue?

Abrupt climate change has marked the earth's history for eons. Ice cores from Greenland, for instance, reveal that wild temperature swings (*opposite top*) punctuated the gradual warming that brought the planet out of the last ice age starting about 18,000 years ago. Fossil shells in lake sediments from Mexico's Yucatan Peninsula record sudden and severe droughts (*opposite bottom*) because a diagnostic ratio of oxygen isotopes in the shells shoots up when more water evaporates from the lake than falls as rain. Societies have often suffered as a result of these rapid shifts.

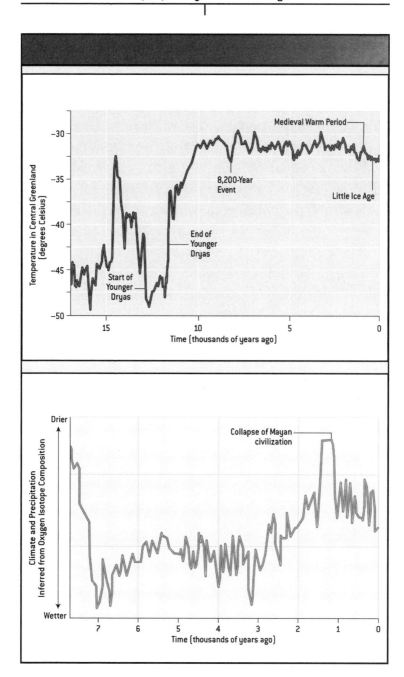

Point of No Return

Be they warm spells, cold snaps or prolonged droughts, the precipitous climate changes of the past all happened for essentially the same reason. In each case, a gradual change in temperature or other physical condition pushed a key driver of climate toward an invisible threshold. At the point that threshold was crossed, the climate driver—and thus the climate as well—flipped to a new and different state and usually stayed there for a long time.

Crossing a climate threshold is similar to flipping a canoe. If you are sitting in a canoe on a lake and you lean gradually to one side, the canoe tips, too. You are pushing the canoe toward a threshold—the position after which the boat can no longer stay upright. Lean a bit too far, and the canoe overturns.

Threshold crossings caused history's most extreme climate flips—and point to areas of particular concern for the future. To explain the icy spells recorded in Greenland's ice cores, for example, most scientists implicate altered behavior of currents in the North Atlantic, which are a dominant factor in that region's long-term weather patterns.

Eastern North America and Europe enjoy temperate conditions (like today's) when salty Atlantic waters warmed by southern sunshine flow northward across the equator. During far northern winters, the salty water arriving from the south becomes cold and dense enough to sink east and west of Greenland, after which

Crossing the Threshold

Global warming alters ambient conditions little by little. But even this kind of slow, steady change can push climate drivers, such as well-established ocean currents or patterns of rainfall, to a critical point at which they lurch abruptly into a new and different state. That switch brings with it an associated shift in climate—with potentially challenging consequences to people and societies. Once a climate driver crosses its so-called threshold, the changes that ensue could persist for millennia. Many thresholds may still await discovery; here are three that scientists have identified:

1. **Climate driver:** Ocean currents in the North Atlantic carry warmth northward from tropics, keeping western Europe's winters mild.
   Threshold crossing: Freshening of surface waters in the far north slows down these currents, possibly stopping them altogether.
   Resulting climate shift: Temperatures plummet in the region, and climate in Europe and the eastern U.S. becomes more like Alaska's.
   Social consequences: Agriculture suffers in regions around the world, and key navigation routes become clogged with ice.

2. **Climate driver:** Rainwater that is recycled through plants (absorbed by their roots and

continued on following page

continued from previous page

returned to the air through evaporation from their leaves) provides much of the precipitation in the world's grain belts.

Threshold crossing: A minor dry spell wilts or kills too many plants, and recycled rainfall disappears, reinforcing the drying in a vicious cycle.

Resulting climate shift: A potentially mild dry spell is enhanced and prolonged into a severe drought.

Social consequences: Parched land can no longer support crops; famine strikes those who cannot trade for the remaining grain in the world market.

3. **Climate driver:** Currents in the Pacific Ocean determine major patterns of sea-surface temperature, which in turn control regional weather patterns.

Threshold crossing: Natural phenomena, such as El Niño, cause subtle changes in sea-surface temperatures, although scientists are still not sure why.

Resulting climate shift: Weather patterns on adjacent continents shift, triggering severe storms or droughts where they typically do not occur.

Social consequences: Some croplands dry up while other places incur damage from intense storms.

it migrates southward along the seafloor. Meanwhile, as the cooled water sinks, warm currents from the south flow northward to take its place. The sinking water thereby drives what is called a conveyor belt circulation that warms the north and cools the south.

Ice cores contain evidence that sudden cold periods occurred after the North Atlantic became less salty, perhaps because freshwater lakes burst through the walls of glaciers and found their way to the sea. Researchers identify this rush of freshwater as the first phase of a critical threshold crossing because they know freshening the North Atlantic can slow or shut off the conveyor, shifting climate as a result.

Diluted by water from the land, seawater flowing in from the south would become less salty and thus

Melting Toward a Cold Snap?

As global warming continues to heat up the planet, many scientists fear that large pulses of freshwater melting off the Greenland ice sheet and other frozen northern landmasses could obstruct the so-called North Atlantic conveyor, the system of ocean currents that brings warmth to Europe and strongly influences climate elsewhere in the world. A conveyor shutdown—or even a significant slowdown—could cool the North Atlantic region even as global temperatures continue to rise. Other challenging and abrupt climate changes would almost certainly result.

continued on following page

continued from previous page

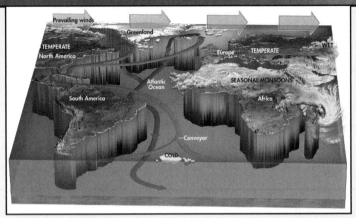

Conveyor On

Salty ocean currents [*dark gray*] flowing northward from the tropics warm prevailing winds [*large arrows*] as they blow eastward toward Europe. The heat-bearing currents, which are dense, become even denser as they lose heat to the atmosphere. Eventually the cold, salty water becomes dense enough to sink near Greenland. It then migrates southward along the seafloor [*light gray*], leaving a void that draws more warm water from the south to take its place.

Resulting Climate

When the North Atlantic conveyor is active, temperate conditions with relatively warm winters enable rich agricultural production in much of Europe and North America. Seasonal monsoons fuel growing seasons in broad swaths of Africa and the Far East. Central Asia is wet, and Antarctica and the South Atlantic are typically cold.

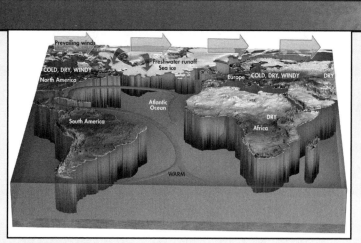

Conveyor Off

If too much freshwater enters the North Atlantic, it dilutes the salty currents from the south. Surface waters no longer become dense enough to sink, no matter how cold the water gets, and the conveyor shuts down or slows. Prevailing winds now carry frigid air eastward [*large arrows*]. This cold trend could endure for decades or more—until southern waters become salty enough to overwhelm the fresher water up north, restarting the conveyor in an enormous rush.

Resulting Climate

As the conveyor grows quiet, winters become harsher in much of Europe and North America, and agriculture suffers. These regions, along with those that usually rely on seasonal monsoons, suffer from droughts sometimes enhanced by stronger winds. Central Asia gets drier, and many regions in the Southern Hemisphere become warmer than usual.

less dense, possibly to the point that it could freeze into sea ice before it had a chance to sink. With sinking stopped and the conveyor halted, rain and snow that fell in the north could not be swept into the deep ocean and carried away. Instead they would accumulate on the sea surface and freshen the North Atlantic even more. The conveyor then would stay quiet, leaving nearby continents with climates more like Siberia's.

Chilling Warmth

Eight thousand years have passed since the last of the biggest North Atlantic cold snaps. Could it be that humans are actually "leaning" in the right direction to avoid flipping the climate's canoe? Perhaps, but most climate experts suspect instead that we are rocking the boat—by changing so many aspects of our world so rapidly. Particularly worrisome are human-induced increases in atmospheric concentrations of greenhouse gases, which are promoting global warming [see "Defusing the Global Warming Time Bomb," by James Hansen; SCIENTIFIC AMERICAN, March; www.sciam.com/ontheweb].

The United Nations-sanctioned Intergovernmental Panel on Climate Change has predicted that average global temperatures will rise 1.5 to 4.5 degrees C in the next 100 years. Many computer models that agree with this assessment also predict a slowdown of the North Atlantic conveyor. (As ironic as it may sound, gradual warming could lead to a sudden cooling of many degrees.) Uncertainties abound, and although a new ice age is not thought credible, the resulting changes could be notably larger than they were during the Little Ice Age, when the Thames in London froze and glaciers rumbled down the Alps.

Perhaps of greater concern than cold spells up north are the adverse effects that would probably strike other parts of the world concurrently. Records of climate across broad areas of Africa and Asia that typically benefit from a season of heavy monsoons indicate that these areas were particularly dry whenever the North Atlantic region was colder than the lands around it. Even the cooling from a conveyor slowdown might be enough to produce the drying. With billions of people relying on monsoons to water crops, even a minor drought could lead to widespread famine.

The consequences of future North Atlantic freshening and cooling may make life more difficult even for people living in regions outside the extreme cold or drought. Unease over such broad impacts spurred the U.S. Department of Defense to request that a think tank called the Global Business Network assess the possible national security implications of a total shutdown of the North Atlantic conveyor. Many scientists, including me, think that a moderate slowdown is much more likely than a total shutdown; either way, the seriousness of the potential outcome makes considering the worst-case implications worthwhile. As the Global Business Network report states, "Tensions could mount around the world . . . Nations with the resources to do so may build virtual fortresses around their countries, preserving resources for themselves. Less fortunate nations . . . may initiate in struggles for access to food, clean water, or energy."

Floods and Droughts

Even if a slowdown of the North Atlantic conveyor never happens, global warming could bring about troubling threshold crossings elsewhere. The grain belts that stretch across the interiors of many midlatitude continents face a regional risk of prolonged drought. Most climate models produce greater summertime drying over these areas as average global temperatures rise, regardless of what happens in the North Atlantic. The same forecasts suggest that greenhouse-induced warming will increase rainfall overall, possibly in the form of more severe storms and flooding; however, those events—significant problems on their own—are not expected to offset the droughts.

Summer drying could cause a relatively mild drought to worsen and persist for decades or more. This transition would occur because of a vulnerability of the grain belts: for precipitation, they rely heavily on rainfall that local plants recycle rather than on new moisture blown in from elsewhere. The plants' roots normally absorb water that would otherwise soak through the ground to streams and flow to the sea. Some of that water then returns to the air by evaporating through their leaves. As the region begins to suffer drier summers, however, the plants wilt and possibly die, thereby putting less water back into the air. The vital threshold is crossed when the plant population shrinks to the point that the recycled rainfall becomes too meager to sustain the population. At that point more plants die, and the rainfall diminishes further—in a vicious cycle like the one that turned the Sahara into a desert 5,000 years ago. The region has shown no signs of greening ever since.

Scientists fear they have not yet identified many of the thresholds that, when crossed, would lead to changes in regional climates. That knowledge gap is worrisome, because humans could well be doing many things to tip the climate balance in ways we will regret. Dancing in a canoe is not usually recommended, yet dance we do: We are replacing forests with croplands, which increases how much sunlight the land reflects; we are pumping water out of the ground, which changes how much water rivers carry to the oceans; and we are altering the quantities of trace gases and particulates in the atmosphere, which modifies the characteristics of clouds, rainfall and more.

Facing the Future

Negative consequences of a major climate shift can be mitigated if the change occurs gradually or is expected. Farmers anticipating a drought can drill wells, or learn to grow crops less dependent on water, or simply cut their losses and move elsewhere. But unexpected change can be devastating. A single, surprise drought year may at first bankrupt or starve only the most marginal farmers, but damage worsens as the drought lengthens—especially if no one had time to prepare. Unfortunately, scientists have little ability to predict when abrupt climate change will occur and what form it will take.

Despite the potentially enormous consequences of a sudden climate transformation, the vast majority of climate research and policymaking has addressed gradual shifts—most notably by calling for global reductions of carbon emissions as a way to slow the gradual rise in global temperatures. Although such reductions would probably help limit climate instability, thought should also be given specifically to avoiding abrupt changes. At one extreme, we might decide to ignore the prospect altogether and hope that nothing happens or that we are able to deal with whatever does happen; business-as-usual did sink the *Titanic*, but many other unprepared ships have crossed the North Atlantic unscathed. On the other hand, we might seriously alter our behavior to keep the human effects on climate small enough to make a catastrophic shift less likely. Curbing global warming would be a step in the right direction. Further investigation of climate thresholds and their vulnerabilities to human activities should illuminate other useful actions.

A third strategy would be for societies to shore up their abilities to cope with abrupt climate change before the next surprise is upon us, as suggested by the U.S. National Research Council. The authors of the council's report pointed out that some former societies have bent in response to climate change when others have broken. The Viking settlers in Greenland abandoned their weakening settlement as the onset of the Little Ice Age made their way of life marginal or unsustainable, while their neighbors, the Thule Inuit, survived. Understanding what separates bending from breaking could prove constructive. Plans designed to help ease difficulties if a crisis develops could be made at little or no cost. Communities could plant trees now to help hold soil during the next windy dry spell, for example, and they could agree now on who will have access to which water supplies when that resource becomes less abundant.

For now, it appears likely that humans are rocking the boat, pushing certain aspects of climate closer to the thresholds that could unleash sudden changes. Such events would not trigger a new ice age or otherwise rival the fertile imaginations of the writers of the silver screen, but they could pose daunting challenges for humans and other living things on earth. It is well worth considering how societies might increase their resiliency to the potential consequences of an abrupt shift—or even how to avoid flipping the climate canoe in the first place.

The Author

Richard B. Alley is professor of geosciences at Pennsylvania State University and an associate at the Earth System Science Center there. Since earning his Ph.D. in geology from the University of Wisconsin-Madison in 1987, Alley has focused his work on understanding how glaciers and ice sheets record climate change, how their flow affects sea level, and how they erode and deposit sediment. Alley has spent three field seasons in Antarctica and five in Greenland. During those visits, he helped to harvest many of the key ice cores that are revealing the history of the earth's climate over the past 110,000 years. In 2001 Alley won the national Phi Beta Kappa Science Award for his nontechnical book The Two-Mile Time Machine, *which details the results from a two-mile-long ice core from Greenland.*

In the early 1990s, clad in down suits and other below-zero protective gear, scientists drilled cores two miles deep into the frozen ice covering Greenland and pulled up hundreds of thousands of years worth of climate history. As one season's snow accumulates atop another's, the older snow is gradually compressed into ice. From bubbles of air frozen in this ice, scientists can extract information about global temperatures and their shifts over vast stretches of time.

Ice cores have now been drilled around the world, and scientists have used them to piece together climate histories that go back hundreds of thousands of years. As a result, they have made some important discoveries about the close relationship between carbon dioxide levels *and temperature, how climate differs in various parts of the world, and how quickly local climate can change. This article, published in the early days of ice-core analysis, describes the researchers' discovery that global temperatures can swing quickly, a phenomenon that is now widely understood to be a cornerstone of the earth's climate history. —KH*

"Core Questions"
by Marguerite Holloway
Scientific American, December 1993

Discussion of whether human economic activity can affect the climate has generally rested on a comforting assumption: if change did occur, it would occur gradually. There would be time to respond. That assumption has been made untenable by analysis of glacial ice as well as of sediments from the ocean floor.

The findings reveal that far from being stable, the earth's climate has always changed quite abruptly— both during times of glaciation and, as the newest studies indicate, during interglacial periods such as the current one. "These are very exciting results," says M. Granger Morgan of Carnegie Mellon University. "Changes took place over shorter time scales than people had expected."

The ice core results come from the efforts of two teams of researchers, one European and one American. Five years ago they set out to find information that would create a more complete record of climates past. The groups selected locations 19 miles apart on the Greenland ice sheet. The European team finished their work in 1992; the Americans finished this summer. The cores they drilled and recovered sampled the ice and snow to a depth of 10,013 feet, encompassing a 250,000-year span of climatological history. (In another venture, Russian scientists are drilling in the Vostok site on the Antarctic ice sheet, hoping to capture at least 500,000 years of evidence.)

The unexpected picture of climate that has emerged from these cores has been reported in a series of articles in *Nature*. The European group—the Greenland Ice-Core Project, or GRIP—found that the last interglacial period, called the Eemian, was characterized by the sudden onset of cold periods that lasted for decades or centuries.

Although such vacillations had been observed in data from glacial times, finding them in the Eemian is significant because the climate was, on average, only a few degrees warmer than it is now. The studies "have concluded that 130,000 years ago, when the earth was as warm as it is today, there were very rapid changes from warm to cold climates," explains Michael L. Bender of the University of Rhode Island. "If that conclusion stands up, it is going to be extremely important because the very stable climate that the earth has had for the past 10,000 years will not necessarily stay that way."

The U.S. team—the Greenland Ice Sheet Project II, or GISP2—finished drilling six months after the

Europeans. GISP2 has also documented rapid and dramatic climatic fluctuations. "That is the importance of having two cores," remarks Scott J. Lehman of the Woods Hole Oceanographic Institution. Because of compression and different characteristics of ice, "it'll be interesting to see any corroboration," Lehman says.

Marine sediment cores, taken by oceanographers, have supported the GRIP and GISP2 findings. By examining the presence of plankton that thrive at various temperatures, Lehman, Gerard Bond of Columbia University's Lamont-Doherty Earth Observatory and other researchers have been able to chart changes in temperature in the North Atlantic. The various groups have reported that the temperature record of the ice is echoed by the seafloor, suggesting that there are links between the temperature changes in the ocean and in the atmosphere.

The factors that cause the abrupt changes remain obscure. One theory holds that the heat-carrying capacity of the Atlantic Ocean—described as a conveyor belt—is somehow altered as fresh water is released by melting ice. These changes cause and are caused by changes in climate. Another hypothesis suggests that the conveyor is disrupted by global variations in rainfall.

For now, climate modeling is likely to offer only limited help in clarifying the reasons for the dynamic change. Although most models have found that doubling of carbon dioxide will result in a global temperature increase of 1.5 to 4.5 degrees Celsius, they are far from being able to incorporate all aspects of the climate system. "We are barely able to model the oceans; we cannot yet couple them with atmospheric models," Lehman says. Without a good model of these interactions, "the possibility of sudden changes is explicitly not allowed." Lehman goes on to note that one or two models have tried to include both elements: "And what do we get in them? Surprises."

So the Greenland findings, in sphinx-like manner, continue to pose questions. Were the changes local or global? If rapid fluctuations are the norm, why is the contemporary climate so stable? Could the accumulation of greenhouse gases trigger a dramatic, and potentially devastating, oscillation today?

"It is the biggest event this year," says Andrew J. Weaver of the University of Victoria in British Columbia, of the Greenland results. "The fact that interglacials are not times of stable climate," Bond adds, "is a warning that we are poised between modes and could bring on a switch."

Global temperatures have risen and fallen in fairly regular and orderly cycles during the last 300,000 years or so, driven in part by extremely predictable changes in the earth's orbit around the sun. Scientists study the record of these temperature fluctuations in the gases that are preserved in long-frozen bubbles in ice cores.

But something odd happened the last time the earth was supposed to enter a glacial cycle: It didn't.

For a long time, some researchers called that a lucky happenstance, because cultures across the world have thrived in the relative warmth. William F. Ruddiman sees it differently: What if this irregular warming was not a mere blip in the earth's ordinary temperature cycle, but the direct result of human-influenced emission of greenhouse gasses beginning thousands of years ago? What if humans began affecting climate thousands of years before the advent of the Industrial Revolution and the introduction of coal-burning factories and power plants? Ruddiman presents a compelling case in this article, but much of the evidence remains circumstantial. More research needs to be done before we can definitively pinpoint when human activity began to alter global climate and throw it off its normal cycles of orderly and predictable temperature fluctuations. —KH

"How Did Humans First Alter Global Climate?"
by William F. Ruddiman
Scientific American, **March 2005**

The scientific consensus that human actions first began to have a warming effect on the earth's climate within the past century has become part of the public perception as well. With the advent of coal-burning factories and power plants, industrial societies began releasing carbon dioxide (CO_2) and other greenhouse gases into the air. Later, motor vehicles added to such emissions. In this scenario, those of us who have lived during the industrial era are responsible not only for the gas buildup in the atmosphere but also for at least part of the accompanying global warming trend. Now, though, it seems our ancient agrarian ancestors may have begun adding these gases to the atmosphere many millennia ago, thereby altering the earth's climate long before anyone thought.

New evidence suggests that concentrations of CO_2 started rising about 8,000 years ago, even though natural trends indicate they should have been dropping. Some 3,000 years later the same thing happened to methane, another heat-trapping gas. The consequences of these surprising rises have been profound. Without them, current temperatures in northern parts of North America and Europe would be cooler by three to four degrees Celsius—enough to make agriculture difficult. In addition, an incipient ice age—marked by the appearance of small ice caps—would probably have begun several thousand years ago in parts of northeastern Canada. Instead the earth's climate has remained relatively warm and stable in recent millennia.

Until a few years ago, these anomalous reversals in greenhouse gas trends and their resulting effects on climate had escaped notice. But after studying the problem for some time, I realized that about 8,000 years ago the gas trends stopped following the pattern that would be predicted from their past long-term behavior, which had been marked by regular cycles. I concluded that human activities tied to farming—primarily agricultural deforestation and crop irrigation—must have added the extra CO_2 and methane to the atmosphere. These activities explained both the reversals in gas trends and the ongoing increases right up to the start of the industrial era. Since then, modern technological innovations have brought about even faster rises in greenhouse gas concentrations.

My claim that human contributions have been altering the earth's climate for millennia is provocative and controversial. Other scientists have reacted to this proposal with the mixture of enthusiasm and skepticism that is typical when novel ideas are put forward, and testing of this hypothesis is now under way.

The Current View

This new idea builds on decades of advances in understanding long-term climate change. Scientists have known since the 1970s that three predictable variations in the earth's orbit around the sun have exerted the dominant control over long-term global climate for millions of years. As a consequence of these orbital cycles (which operate over 100,000, 41,000 and 22,000 years), the amount of solar radiation reaching various parts of the globe during a given season can differ by more than 10 percent. Over the past three million years, these regular changes in the amount of sunlight reaching the planet's surface have produced a long sequence of ice ages (when great areas of Northern Hemisphere continents were covered with ice) separated by short, warm interglacial periods.

Dozens of these climatic sequences occurred over the millions of years when hominids were slowly evolving toward anatomically modern humans. At the end of the most recent glacial period, the ice sheets that had blanketed northern Europe and North America for the previous 100,000 years shrank and, by 6,000 years ago, had disappeared. Soon after, our ancestors built cities, invented writing and founded religions. Many scientists credit much of the progress of civilization to this naturally warm gap between less favorable glacial intervals, but in my opinion this view is far from the full story.

In recent years, cores of ice drilled in the Antarctic and Greenland ice sheets have provided extremely valuable evidence about the earth's past climate, including changes in the concentrations of the greenhouse gases. A three-kilometer-long ice core retrieved from Vostok Station in Antarctica during the 1990s contained trapped bubbles of ancient air that revealed the composition of the atmosphere (and the gases) at the time the ice layers formed. The Vostok ice confirmed that concentrations of CO_2 and methane rose and fell in a regular pattern during virtually all of the past 400,000 years.

Particularly noteworthy was that these increases and decreases in greenhouse gases occurred at the same intervals as variations in the intensity of solar radiation and the size of the ice sheets. For example, methane concentrations fluctuate mainly at the 22,000-year tempo of an orbital cycle called precession. As the earth spins on its rotation axis, it wobbles like a top, slowly swinging the Northern Hemisphere closer to and then farther from the sun. When this precessional wobble brings the northern continents nearest the sun during the summertime, the atmosphere gets a notable boost of methane from its primary natural source—the decomposition of plant matter in wetlands.

After wetland vegetation flourishes in late summer, it then dies, decays and emits carbon in the form of methane, sometimes called swamp gas. Periods of maximum summertime heating enhance methane production in two primary ways: In southern Asia, the warmth draws additional moisture-laden air in from the Indian Ocean, driving strong tropical monsoons that flood regions that might otherwise stay dry. In far northern Asia and Europe, hot summers thaw boreal wetlands for longer periods of the year. Both processes enable more vegetation to grow, decompose and emit methane every 22,000 years. When the Northern Hemisphere veers farther from the sun, methane emissions start to decline. They bottom out 11,000 years later—the point in the cycle when Northern Hemisphere summers receive the least solar radiation.

Unexpected Reversals

Examining records from the Vostok ice core closely, I spotted something odd about the recent part of the record. Early in previous interglacial intervals, the methane concentration typically reached a peak of almost 700 parts per billion (ppb) as precession brought summer radiation to a maximum. The same thing happened 11,000 years ago, just as the current interglacial period began. Also in agreement with prior cycles, the methane concentration then declined by 100 ppb as summer sunshine subsequently waned. Had the recent trend continued to mimic older interglacial intervals, it would have fallen to a value near 450 ppb during the current minimum in summer heating. Instead the trend reversed direction 5,000 years ago and rose gradually back to almost 700 ppb just before the start of the industrial era. In short, the methane concentration rose when it should have fallen, and it ended up 250 ppb higher than the equivalent point in earlier cycles.

Like methane, CO_2 has behaved unexpectedly over the past several thousand years. Although a complex combination of all three orbital cycles controls CO_2 variations, the trends during previous interglacial intervals were all surprisingly similar to one another. Concentrations peaked at 275 to 300 parts per million (ppm) early in each warm period, even before the last remnants of the great ice sheets finished melting. The CO_2 levels then fell steadily over the next 15,000 years to an average of about 245 ppm. During the current interglacial interval, CO_2 concentrations reached the expected peak around 10,500 years ago and, just as anticipated, began a similar decline. But instead of continuing to drop steadily through modern times, the trend reversed direction 8,000 years ago. By the start of the industrial era, the concentration had risen to 285 ppm—roughly 40 ppm higher than expected from the earlier behavior.

What could explain these unexpected reversals in the natural trends of both methane and CO_2? Other investigators suggested that natural factors in the climate system provided the answer. The methane increase has been ascribed to expansion of wetlands in Arctic regions and the CO_2 rise to natural losses of carbon-rich vegetation on the continents, as well as to changes in the chemistry of the ocean. Yet it struck me that these explanations were doomed to fail for a simple reason. During the four preceding interglaciations, the major factors thought to influence greenhouse gas concentrations in the atmosphere were nearly the same as in recent millennia. The northern ice sheets had melted, northern forests had reoccupied the land uncovered by ice, meltwater from the ice had returned sea level to its high inter-glacial position, and solar radiation driven by the earth's orbit had increased and then begun to decrease in the same way.

Why, then, would the gas concentrations have fallen during the last four interglaciations yet risen only during the current one? I concluded that something new to the natural workings of the climate system must have been operating during the past several thousand years.

The Human Connection

The most plausible "new factor" operating in the climate system during the present interglaciation is farming. The basic timeline of agricultural innovations is well known. Agriculture originated in the Fertile Crescent region of the eastern Mediterranean around 11,000 years ago, shortly thereafter in northern China, and several thousand years later in the Americas. Through subsequent millennia it spread to other regions and increased in sophistication. By 2,000 years ago, every crop food eaten today was being cultivated somewhere in the world.

Several farming activities generate methane. Rice paddies flooded by irrigation generate methane for the same reason that natural wetlands do—vegetation decomposes in the stagnant standing water. Methane is also released as farmers burn grasslands to attract game and promote growth of berries. In addition, people and their domesticated animals emit methane with feces and belches. All these factors probably contributed to a gradual rise in methane as human populations grew slowly, but only one process seems likely to have accounted for the abruptness of the reversal from a natural methane decline to an unexpected rise around 5,000 years ago—the onset of rice irrigation in southern Asia.

Farmers began flooding lowlands near rivers to grow wet-adapted strains of rice around 5,000 years ago in the south of China. With extensive floodplains lying within easy reach of several large rivers, it makes sense that broad swaths of land could have been flooded soon after the technique was discovered, thus explaining the quick shift in the methane trend. Historical records also indicate a steady expansion in rice irrigation throughout the interval when methane values were rising. By 3,000 years ago the technique had spread south into Indochina and west to the Ganges River Valley in India, further increasing methane emissions. After 2,000 years, farmers began to construct rice paddies on the steep hillsides of Southeast Asia.

Future research may provide quantitative estimates of the amount of land irrigated and methane generated through this 5,000-year interval. Such estimates will probably be difficult to come by, however, because repeated irrigation of the same areas into modern times has probably disturbed much of the earlier evidence. For now, my case rests mainly on the basic fact that the methane trend went the "wrong way" and that farmers began to irrigate wetlands at just the right time to explain this wrong-way trend.

Another common practice tied to farming—deforestation—provides a plausible explanation for the start of the anomalous CO_2 trend. Growing crops in naturally forested areas requires cutting trees, and farmers began to clear forests for this purpose in Europe and China by 8,000 years ago, initially with axes made of stone and later from bronze and then iron. Whether the fallen trees were burned or left to rot, their carbon would have soon oxidized and ended up in the atmosphere as CO_2.

Scientists have precisely dated evidence that Europeans began growing nonindigenous crop plants such as wheat, barley and peas in naturally forested areas just as the CO_2 trend reversed 8,000 years ago. Remains of these plants, initially cultivated in the Near East, first appear in lake sediments in southeastern Europe and then spread to the west and north over the next several thousand years. During this interval, silt and clay began to wash into rivers and lakes from denuded hillsides at increasing rates, further attesting to ongoing forest clearance.

The most unequivocal evidence of early and extensive deforestation lies in a unique historical document—the Doomsday Book. This survey of England, ordered by William the Conqueror, reported that 90 percent of the natural forest in lowland, agricultural regions was cleared as of AD 1086. The survey also counted 1.5 million people living in England at the time, indicating that an average density of 10 people per square kilometer was sufficient to eliminate the forests. Because the advanced civilizations of the major river valleys of China and India had reached much higher population densities several thousand years prior, many historical ecologists have concluded that these regions were heavily deforested some two or even three thousand years ago. In summary, Europe and southern Asia had been heavily deforested long before the start of the industrial era, and the clearance process was well under way throughout the time of the unusual CO_2 rise.

An Ice Age Prevented?

If farmers were responsible for greenhouse gas anomalies this large—250 ppb for methane and 40 ppm for CO_2 by the 1700s—the effect of their practices on the earth's climate would have been substantial. Based on the average sensitivity shown by a range of climate models, the combined effect from these anomalies would have been an average warming of almost 0.8 degree C just before the industrial era. That amount is larger than the 0.6 degree C warming measured during the past century—implying that the effect of early farming on climate rivals or even exceeds the combined changes registered during the time of rapid industrialization.

How did this dramatic warming effect escape recognition for so long? The main reason is that it was masked by natural climatic changes in the opposite direction. The earth's orbital cycles were driving a simultaneous natural cooling trend, especially at high northern latitudes. The net temperature change was a gradual summer cooling trend lasting until the 1800s.

Had greenhouse gases been allowed to follow their natural tendency to decline, the resulting cooling would have augmented the one being driven by the drop in summer radiation, and this planet would have become considerably cooler than it is now. To explore this possibility, I joined with Stephen J. Vavrus and John E. Kutzbach of the University of Wisconsin–Madison to use a climate model to predict modern-day temperature in the absence of all human-generated greenhouse gases.

The model simulates the average state of the earth's climate—including temperature and precipitation—in response to different initial conditions.

For our experiment, we reduced the greenhouse gas levels in the atmosphere to the values they would have reached today without early farming or industrial emissions. The resulting simulation showed that our planet would be almost two degrees C cooler than it is now—a significant difference. In comparison, the global mean temperature at the last glacial maximum 20,000 years ago was only five to six degrees C colder than it is today. In effect, current temperatures would be well on the way toward typical glacial temperatures had it not been for the greenhouse gas contributions from early farming practices and later industrialization.

I had also initially proposed that new ice sheets might have begun to form in the far north if this natural cooling had been allowed to proceed. Other researchers had shown previously that parts of far northeastern Canada might be ice covered today if the world were cooler by just 1.5 to two degrees C—the same amount of cooling that our experiment suggested has been offset by the greenhouse gas anomalies. The later modeling effort with my Wisconsin colleagues showed that snow would now persist into late summer in two areas of northeastern Canada: Baffin Island, just east of the mainland, and Labrador, farther south. Because any snow that survives throughout the summer will accumulate in thicker piles year by year and eventually become glacial ice, these results suggest that a new ice age would have begun in northeast Canada several millennia ago, at least on a small scale.

This conclusion is startlingly different from the traditional view that human civilization blossomed within a period of warmth that nature provided. As I see it, nature would have cooled the earth's climate, but our ancestors kept it warm by discovering agriculture.

Implications for the Future

The conclusion that humans prevented a cooling and arguably stopped the initial stage of a glacial cycle bears directly on a long-running dispute over what global climate has in store for us in the near future. Part of the reason that policymakers had trouble embracing the initial predictions of global warming in the 1980s was that a number of scientists had spent the previous decade telling everyone almost exactly the opposite—that an ice age was on its way. Based on the new confirmation that orbital variations control the growth and decay of ice sheets, some scientists studying these longer-scale changes had reasonably concluded that the next ice age might be only a few hundred or at most a few thousand years away.

In subsequent years, however, investigators found that greenhouse gas concentrations were rising rapidly and that the earth's climate was warming, at least in part because of the gas increases. This evidence convinced most scientists that the relatively near-term future (the next century or two) would be dominated by global warming rather than by global cooling. This revised prediction, based on an improved understanding of the climate system, led some policymakers to discount all forecasts—whether of global warming or an impending ice age—as untrustworthy.

My findings add a new wrinkle to each scenario. If anything, such forecasts of an "impending" ice age were actually understated: new ice sheets should have begun to grow several millennia ago. The ice failed to grow because human-induced global warming actually began far earlier than previously thought—well before the industrial era.

In these kinds of hotly contested topics that touch on public policy, scientific results are often used for opposing ends. Global-warming skeptics could cite my work as evidence that human-generated greenhouse gases played a beneficial role for several thousand years by keeping the earth's climate more hospitable than it would otherwise have been. Others might counter that if so few humans with relatively primitive technologies were able to alter the course of climate so significantly, then we have reason to be concerned about the current rise of greenhouse gases to unparalleled concentrations at unprecedented rates.

The rapid warming of the past century is probably destined to persist for at least 200 years, until the economically accessible fossil fuels become scarce. Once that happens, the earth's climate should begin to cool gradually as the deep ocean slowly absorbs the pulse of excess CO_2 from human activities. Whether global climate will cool enough to produce the long-

overdue glaciation or remain warm enough to avoid
that fate is impossible to predict.

The Author

*William F. Ruddiman is a marine geologist and professor
emeritus of environmental sciences at the University of
Virginia. He joined the faculty there in 1991 and served
as department chair from 1993 to 1996. Ruddiman first
began studying records of climate change in ocean sediments as a graduate student at Columbia University,
where he received his doctorate in 1969. He then worked
as a senior scientist and oceanographer with the U.S.
Naval Oceanographic Office in Maryland and later as a
senior research scientist at Columbia's Lamont-Doherty
Earth Observatory.*

*The chemistry discussed in the following article
is not as complicated as it first appears, and the
final message is simple: geological history tells us
that the earth's climate has changed dramatically
and quickly in the past, and therefore some life-
altering shifts are quite possibly in our future.
The authors, Paul F. Hoffman and Daniel P. Schrag,
began their investigation into extreme climate
change with strange geological deposits in
Namibia and eventually arrived at a startling
and far-ranging theory that, they propose,*

*explains no less than the Cambrian explosion—
the sudden diversification of animal life-forms
600 million years ago. Their idea is not universally accepted, but it does seemingly account
for a surprising number of geological mysteries
that have thus far defied easy explanation.
Fundamentally, the work depends on the
researchers' intimate understanding of feedback
cycles, both positive and negative. —KH*

"Snowball Earth"
by Paul F. Hoffman and Daniel P. Schrag
Scientific American, January 2000

*Some say the world will end in fire,
Some say in ice.
From what I've tasted of desire
I hold with those who favor fire.
But if it had to perish twice,
I think I know enough of hate
To say that for destruction ice
Is also great
And would suffice.*

—Robert Frost, Fire and Ice (1923)

Our human ancestors had it rough. Saber-toothed cats
and woolly mammoths may have been day-to-day
concerns, but harsh climate was a consuming long-term
challenge. During the past million years, they faced
one ice age after another. At the height of the last icy

episode, 20,000 years ago, glaciers more than two kilometers thick gripped much of North America and Europe. The chill delivered ice as far south as New York City.

Dramatic as it may seem, this extreme climate change pales in comparison to the catastrophic events that some of our earliest microscopic ancestors endured around 600 million years ago. Just before the appearance of recognizable animal life, in a time period known as the Neoproterozoic, an ice age prevailed with such intensity that even the tropics froze over.

Imagine the earth hurtling through space like a cosmic snowball for 10 million years or more. Heat escaping from the molten core prevents the oceans from freezing to the bottom, but ice grows a kilometer thick in the −50 degree Celsius cold. All but a tiny fraction of the planet's primitive organisms die. Aside from grinding glaciers and groaning sea ice, the only stir comes from a smattering of volcanoes forcing their hot heads above the frigid surface. Although it seems the planet might never wake from its cryogenic slumber, the volcanoes slowly manufacture an escape from the chill: carbon dioxide.

With the chemical cycles that normally consume carbon dioxide halted by the frost, the gas accumulates to record levels. The heat-trapping capacity of carbon dioxide—a greenhouse gas—warms the planet and begins to melt the ice. The thaw takes only a few hundred years, but a new problem arises in the meantime: a brutal greenhouse effect. Any creatures that survived the icehouse must now endure a hothouse.

As improbable as it may sound, we see clear evidence that this striking climate reversal—the most extreme imaginable on this planet—happened as many as four times between 750 million and 580 million years ago. Scientists long presumed that the earth's climate was never so severe; such intense climate change has been more widely accepted for other planets such as Venus [see "Global Climate Change on Venus," by Mark A. Bullock and David H. Grinspoon; SCIENTIFIC AMERICAN, March 1999]. Hints of a harsh past on the earth began cropping up in the early 1960s, but we and our colleagues have found new evidence in the past eight years that has helped us weave a more explicit tale that is capturing the attention of geologists, biologists and climatologists alike.

Thick layers of ancient rock hold the only clues to the climate of the Neoproterozoic. For decades, many of those clues appeared rife with contradiction. The first paradox was the occurrence of glacial debris near sea level in the tropics. Glaciers near the equator today survive only at 5,000 meters above sea level or higher, and at the worst of the last ice age they reached no lower than 4,000 meters. Mixed in with the glacial debris are unusual deposits of iron-rich rock. These deposits should have been able to form only if the Neoproterozoic oceans and atmosphere contained little or no oxygen, but by that time the atmosphere had already evolved to nearly the same mixture of gases as it has today. To confound matters, rocks known to form in warm water seem to have accumulated just after the glaciers receded. If the earth were ever cold enough to ice over completely, how did it warm up again? In addition, the carbon isotopic signature in the rocks hinted at a prolonged drop in biological productivity. What could have caused this dramatic loss of life?

Each of these long-standing enigmas suddenly makes sense when we look at them as key plot developments in the tale of a "snowball earth." The theory has garnered cautious support in the scientific community since we first introduced the idea in the journal *Science* a year and a half ago. If we turn out to be right, the tale does more than explain the mysteries of Neoproterozoic climate and challenge long-held assumptions about the limits of global change. These extreme glaciations occurred just before a rapid diversification of multi-cellular life, culminating in the so-called Cambrian explosion between 575 and 525 million years ago. Ironically, the long periods of isolation and extreme environments on a snowball earth would most likely have spurred on genetic change and could help account for this evolutionary burst.

The search for the surprisingly strong evidence for these climatic events has taken us around the world. Although we are now examining Neoproterozoic rocks in Australia, China, the western U.S. and the Arctic islands of Svalbard, we began our investigations in 1992 along the rocky cliffs of Namibia's Skeleton

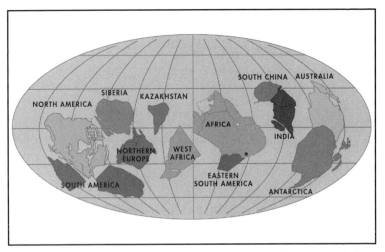

Earth's landmasses were most likely clustered near the equator during the global glaciations that took place around 600 million years ago. Although the continents have since shifted position, relics of the debris left behind when the ice melted are exposed at dozens of points on the present land surface, including what is now Namibia (*gray dot*).

Coast. In Neoproterozoic times, this region of southwestern Africa was part of a vast, gently subsiding continental shelf located in low southern latitudes.

There we see evidence of glaciers in rocks formed from deposits of dirt and debris left behind when the ice melted. Rocks dominated by calcium- and magnesium-carbonate minerals lie just above the glacial debris and harbor the chemical evidence of the hothouse that followed. After hundreds of millions of years of burial, these now exposed rocks tell the story that scientists first began to piece together 35 years ago.

In 1964 W. Brian Harland of the University of Cambridge pointed out that glacial deposits dot

Neoproterozoic rock outcrops across virtually every continent. By the early 1960s scientists had begun to accept the idea of plate tectonics, which describes how the planet's thin, rocky skin is broken into giant pieces that move atop a churning mass of hotter rock below. Harland suspected that the continents had clustered together near the equator in the Neoproterozoic, based on the magnetic orientation of tiny mineral grains in the glacial rocks. Before the rocks hardened, these grains aligned themselves with the magnetic field and dipped only slightly relative to horizontal because of their position near the equator. (If they had formed near the poles, their magnetic orientation would be nearly vertical.)

Realizing that the glaciers must have covered the tropics, Harland became the first geologist to suggest that the earth had experienced a great Neoproterozoic ice age [see "The Great Infra-Cambrian Glaciation," by W. B. Harland and M. J. S. Rudwick; SCIENTIFIC AMERICAN, August 1964]. Although some of Harland's contemporaries were skeptical about the reliability of the magnetic data, other scientists have since shown that Harland's hunch was correct. But no one was able to find an explanation for how glaciers could have survived the tropical heat.

At the time Harland was announcing his ideas about Neoproterozoic glaciers, physicists were developing the first mathematical models of the earth's climate. Mikhail Budyko of the Leningrad Geophysical Observatory found a way to explain tropical glaciers using equations that describe the way solar radiation interacts with the earth's surface and atmosphere to control climate. Some geographic surfaces reflect more of the sun's incoming energy than others, a quantifiable characteristic known as albedo. White snow reflects the most solar energy and has a high albedo, darker-colored seawater has a low albedo, and land surfaces have intermediate values that depend on the types and distribution of vegetation.

The more radiation the planet reflects, the cooler the temperature. With their high albedo, snow and ice cool the atmosphere and thus stabilize their own existence. Budyko knew that this phenomenon, called the ice-albedo feedback, helps modern polar ice sheets to grow. But his climate simulations also revealed that this feedback can run out of control. When ice formed at latitudes lower than around 30 degrees north or south of the equator, the planet's albedo began to rise at a faster rate because direct sunlight was striking a larger surface area of ice per degree of latitude. The feedback became so strong in his simulation that surface temperatures plummeted and the entire planet froze over.

Frozen and Fried

Budyko's simulation ignited interest in the fledgling science of climate modeling, but even he did not believe the earth could have actually experienced a runaway freeze. Almost everyone assumed that such a catastrophe would have extinguished all life, and yet signs of microscopic algae in rocks up to one billion years old closely resemble modern forms and imply a continuity of life. Also, once the earth had entered a deep freeze, the high albedo of its icy veneer would have driven surface temperatures so low that it seemed there would have been no means of escape. Had such a glaciation occurred, Budyko and others reasoned, it would have been permanent.

The first of these objections began to fade in the late 1970s with the discovery of remarkable communities of organisms living in places once thought too harsh to harbor life. Seafloor hot springs support microbes that thrive on chemicals rather than sunlight. The kind of volcanic activity that feeds the hot springs would have continued unabated in a snowball earth. Survival prospects seem even rosier for psychrophilic, or cold-loving, organisms of the kind living today in the intensely cold and dry mountain valleys of East Antarctica. Cyanobacteria and certain kinds of algae occupy habitats such as snow, porous rock and the surfaces of dust particles encased in floating ice.

The key to the second problem—reversing the runaway freeze—is carbon dioxide. In a span as short as a human lifetime, the amount of carbon dioxide in the atmosphere can change as plants consume the gas for photosynthesis and as animals breathe it out during respiration. Moreover, human activities such as burning fossil fuels have rapidly loaded the air with carbon dioxide since the beginning of the Industrial Revolution in the late 1700s. In the earth's lifetime, however, these carbon sources and sinks become irrelevant compared with geologic processes.

Carbon dioxide is one of several gases emitted from volcanoes. Normally this endless supply of carbon is offset by the erosion of silicate rocks: The chemical breakdown of the rocks converts carbon dioxide to bicarbonate, which is washed to the oceans. There bicarbonate combines with calcium and magnesium ions to produce carbonate sediments, which store a great deal of carbon [see "Modeling the Geochemical Carbon Cycle," by R. A. Berner and A. C. Lasaga; SCIENTIFIC AMERICAN, March 1989].

In 1992 Joseph L. Kirschvink, a geobiologist at the California Institute of Technology, pointed out that during a global glaciation, an event he termed a snowball earth, shifting tectonic plates would continue to build volcanoes and to supply the atmosphere with carbon dioxide. At the same time, the liquid water needed to erode rocks and bury the carbon would be trapped in ice. With nowhere to go, carbon dioxide would collect to incredibly high levels—high enough, Kirschvink proposed, to heat the planet and end the global freeze.

Kirschvink had originally promoted the idea of a Neoproterozoic deep freeze in part because of mysterious iron deposits found mixed with the glacial debris. These rare deposits are found much earlier in earth history when the oceans (and atmosphere) contained very little oxygen and iron could readily dissolve. (Iron is virtually insoluble in the presence of oxygen.)

Kirschvink reasoned that millions of years of ice cover would deprive the oceans of oxygen, so that dissolved iron expelled from seafloor hot springs could accumulate in the water. Once a carbon dioxide-induced greenhouse effect began melting the ice, oxygen would again mix with the seawater and force the iron to precipitate out with the debris once carried by the sea ice and glaciers.

With this greenhouse scenario in mind, climate modelers Kenneth Caldeira of Lawrence Livermore National Laboratory and James F. Kasting of Pennsylvania State University estimated in 1992 that overcoming the runaway freeze would require roughly 350 times the present-day concentration of carbon dioxide. Assuming volcanoes of the Neoproterozoic belched out gases at the same rate as they do today, the planet would have remained locked in ice for up to tens of millions of years before enough carbon dioxide could accumulate to begin melting the sea ice. A snowball earth would be not only the most severe conceivable ice age, it would be the most prolonged.

Carbonate Clues

Kirschvink was unaware of two emerging lines of evidence that would strongly support his snowball earth hypothesis. The first is that the Neoproterozoic glacial deposits are almost everywhere blanketed by carbonate rocks. Such rocks typically form in warm, shallow seas, such as the Bahama Banks in what is now the Atlantic Ocean. If the ice and warm water

continued on page 64

Evolution of a Snowball Earth Event . . .

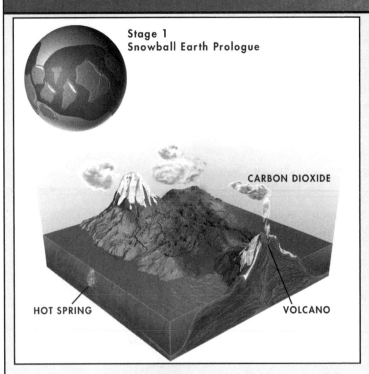

Stage 1
Snowball Earth Prologue

CARBON DIOXIDE

HOT SPRING

VOLCANO

Breakup of a single landmass 770 million years ago leaves small continents scattered near the equator. Formerly landlocked areas are now closer to oceanic sources of moisture. Increased rainfall scrubs more heat-trapping carbon dioxide out of the air and erodes continental rocks more quickly. Consequently, global temperatures fall, and large ice packs form in the polar oceans. The white ice reflects more solar energy than does darker seawater, driving temperatures even lower. This feedback cycle triggers an unstoppable cooling effect that will engulf the planet in ice within a millennium.

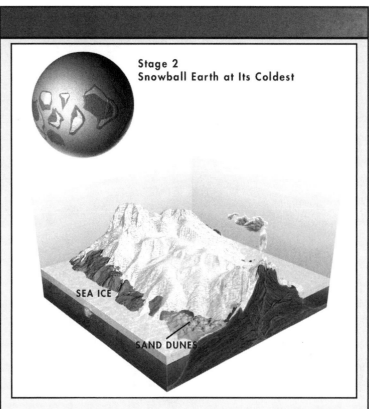

Stage 2
Snowball Earth at Its Coldest

SEA ICE

SAND DUNES

Average global temperatures plummet to −50 degrees Celsius shortly after the runaway freeze begins. The oceans ice over to an average depth of more than a kilometer, limited only by heat emanating slowly from the earth's interior. Most microscopic marine organisms die, but a few cling to life around volcanic hot springs. The cold, dry air arrests the growth of land glaciers, creating vast deserts of windblown sand. With no rainfall, carbon dioxide emitted from volcanoes is not removed from the atmosphere. As carbon dioxide accumulates, the planet warms and sea ice slowly thins.

continued on following page

continued from previous page

. . . And Its Hothouse Aftermath

Stage 3
Snowball Earth as it Thaws

GLACIERS

Concentrations of carbon dioxide in the atmosphere increase 1,000-fold as a result of some 10 million years of normal volcanic activity. The ongoing greenhouse warming effect pushes temperatures to the melting point at the equator. As the planet heats up, moisture from sea ice sublimating near the equator refreezes at higher elevations and feeds the growth of land glaciers. The open water that eventually forms in the tropics absorbs more solar energy and initiates a faster rise in global temperatures. In a matter of centuries, a brutally hot, wet world will supplant the deep freeze.

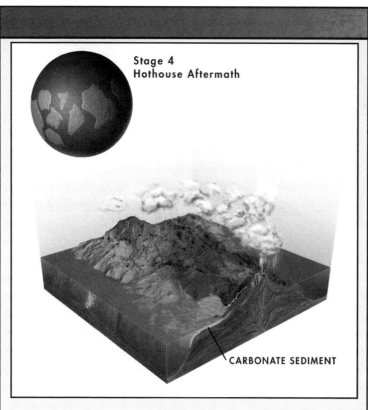

Stage 4
Hothouse Aftermath

CARBONATE SEDIMENT

As tropical oceans thaw, seawater evaporates and works along with carbon dioxide to produce even more intense greenhouse conditions. Surface temperatures soar to more than 50 degrees Celsius, driving an intense cycle of evaporation and rainfall. Torrents of carbonic acid rain erode the rock debris left in the wake of the retreating glaciers. Swollen rivers wash bicarbonate and other ions into the oceans, where they form carbonate sediment. New life-forms—engendered by prolonged genetic isolation and selective pressure—populate the world as global climate returns to normal.

continued from page 59

had occurred millions of years apart, no one would have been surprised. But the transition from glacial deposits to these "cap" carbonates is abrupt and lacks evidence that significant time passed between when the glaciers dropped their last loads and when the carbonates formed. Geologists were stumped to explain so sudden a change from glacial to tropical climates.

Pondering our field observations from Namibia, we realized that this change is no paradox. Thick sequences of carbonate rocks are the expected consequence of the extreme greenhouse conditions unique to the transient aftermath of a snowball earth. If the earth froze over, an ultrahigh carbon dioxide atmosphere would be needed to raise temperatures to the melting point at the equator. Once melting begins, low-albedo seawater replaces high-albedo ice and the runaway freeze is reversed [*see illustrations on pp. 60–63*]. The greenhouse atmosphere helps to drive surface temperatures upward to almost 50 degrees C, according to calculations made last summer by climate modeler Raymond T. Pierrehumbert of the University of Chicago.

Resumed evaporation also helps to warm the atmosphere because water vapor is a powerful greenhouse gas, and a swollen reservoir of moisture in the atmosphere would drive an enhanced water cycle. Torrential rain would scrub some of the carbon dioxide out of the air in the form of carbonic acid, which would rapidly erode the rock debris left bare as the glaciers subsided. Chemical erosion products would quickly build up in the ocean water, leading to the precipitation of carbonate sediment that would rapidly accumulate on the seafloor and later become rock. Structures preserved in the Namibian cap carbonates indicate that they accumulated extremely rapidly, perhaps in only a few thousand years. For example, crystals of the mineral aragonite, clusters of which are as tall as a person, could precipitate only from seawater highly saturated in calcium carbonate.

Cap carbonates harbor a second line of evidence that supports Kirschvink's snowball escape scenario. They contain an unusual pattern in the ratio of two isotopes of carbon: common carbon 12 and rare carbon 13, which has an extra neutron in its nucleus. The same patterns are observed in cap carbonates worldwide, but no one thought to interpret them in terms of a snowball earth. Along with Alan Jay Kaufman, an isotope geochemist now at the University of Maryland, and Harvard University graduate student Galen Pippa Halverson, we have discovered that the isotopic variation is consistent over many hundreds of kilometers of exposed rock in northern Namibia.

Carbon dioxide moving into the oceans from volcanoes is about 1 percent carbon 13; the rest is carbon 12. If the formation of carbonate rocks were the only process removing carbon from the oceans, then the rock would have the same fraction of carbon 13 as that which comes out of volcanoes. But the soft tissues of algae and bacteria growing in seawater also use carbon from the water around them, and their photosynthetic machinery prefers carbon 12 to carbon 13. Consequently, the carbon that is left to build carbonate rocks in a life-filled ocean such as we have today has a higher ratio of carbon 13 to carbon 12 than does the carbon fresh out of a volcano.

The carbon isotopes in the Neoproterozoic rocks of Namibia record a different situation. Just before the glacial deposits, the amount of carbon 13 plummets to levels equivalent to the volcanic source, a drop we think records decreasing biological productivity as ice encrusted the oceans at high latitudes and the earth teetered on the edge of a runaway freeze. Once the oceans iced over completely, productivity would have essentially ceased, but no carbon record of this time interval exists because calcium carbonate could not have formed in an ice-covered ocean. This drop in carbon 13 persists through the cap carbonates atop the glacial deposits and then gradually rebounds to higher levels of carbon 13 several hundred meters above, presumably recording the recovery of life at the end of the hothouse period.

Abrupt variation in this carbon isotope record shows up in carbonate rocks that represent other times of mass extinction, but none are as large or as long-lived. Even the meteorite impact that killed off the dinosaurs 65 million years ago did not bring about such a prolonged collapse in biological activity.

Overall, the snowball earth hypothesis explains many extraordinary observations in the geologic record of the Neoproterozoic world: the carbon isotopic variations associated with the glacial deposits, the paradox of cap carbonates, the evidence for long-lived glaciers at sea level in the tropics, and the associated iron deposits. The strength of the hypothesis is that it simultaneously explains all these salient features, none of which had satisfactory independent explanations. What is more, we believe this hypothesis sheds light on the early evolution of animal life.

Survival and Redemption of Life

In the 1960s Martin J. S. Rudwick, working with Brian Harland, proposed that the climate recovery following a huge Neoproterozoic glaciation paved the way for the explosive radiation of multicellular animal life soon thereafter. Eukaryotes—cells that have a membrane-bound nucleus and from which all plants and animals descended—had emerged more than one billion years earlier, but the most complex organisms that had evolved when the first Neoproterozoic glaciation hit were filamentous algae and unicellular protozoa. It has always been a mystery why it took so long for these primitive organisms to diversify into the 11 animal body plans that show up suddenly in the fossil record during the Cambrian explosion [*see illustration on p. 68*].

A series of global freeze-fry events would have imposed an environmental filter on the evolution of life. All extant eukaryotes would thus stem from the survivors of the Neoproterozoic calamity. Some measure of the extent of eukaryotic extinctions may be evident in universal "trees of life." Phylogenetic trees indicate how various groups of organisms evolved from one another, based on their degrees of similarity. These

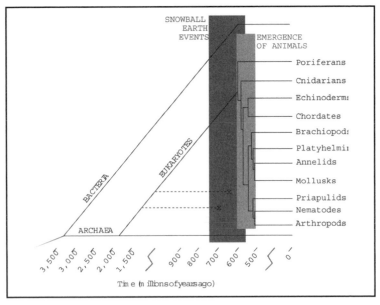

All animals descended from the first eukaryotes, cells with a membrane-bound nucleus, which appeared about two billion years ago. By the time of the first snowball earth episode more than one billion years later, eukaryotes had not developed beyond unicellular protozoa and filamentous algae. But despite the extreme climate, which may have "pruned" the eukaryote tree (*dashed lines*), all 11 animal phyla ever to inhabit the earth emerged within a narrow window of time in the aftermath of the last snowball event. The prolonged genetic isolation and selective pressure intrinsic to a snowball earth could be responsible for this explosion of new life-forms.

days biologists commonly draw these trees by looking at the sequences of nucleic acids in living organisms.

Most such trees depict the eukaryotes' phylogeny as a delayed radiation crowning a long, unbranched stem. The lack of early branching could mean that most eukaryotic lineages were "pruned" during the snowball earth episodes. The creatures that survived the glacial episodes may have taken refuge at hot springs both on the seafloor and near the surface of the ice where photosynthesis could be maintained.

The steep and variable temperature and chemical gradients endemic to ephemeral hot springs would preselect for survival in the hellish aftermath to come. In the face of varying environmental stress, many organisms respond with wholesale genetic alterations. Severe stress encourages a great degree of genetic change in a short time, because organisms that can most quickly alter their genes will have the most opportunities to acquire traits that will help them adapt and proliferate.

Hot-spring communities widely separated geographically on the icy surface of the globe would accumulate genetic diversity over millions of years. When two groups that start off the same are isolated from each other long enough under different conditions, chances are that at some point the extent of genetic mutation will produce a new species. Repopulations occurring after each glaciation would come about under unusual and rapidly changing selective pressures quite different from those preceding the glaciation; such conditions would also favor the emergence of new life-forms.

Martin Rudwick may not have gone far enough with his inference that climatic amelioration following the great Neoproterozoic ice age paved the way for early animal evolution. The extreme climatic events themselves may have played an active role in spawning multicellular animal life.

We have shown how the worldwide glacial deposits and carbonate rocks in the Neoproterozoic record point to an extraordinary type of climatic event, a snowball earth followed by a briefer but equally noxious greenhouse world. But what caused these calamities in the first place, and why has the world been spared such events in more recent history? The first possibility to consider is that the Neoproterozoic sun was weaker by approximately 6 percent, making the earth more susceptible to a global freeze. The slow warming of our sun as it ages might explain why no snowball event has occurred since that time. But convincing geologic evidence suggests that no such glaciations occurred in the billion or so years before the Neoproterozoic, when the sun was even cooler.

The unusual configuration of continents near the equator during Neoproterozoic times may better explain how snowball events get rolling [*see illustration on page 54*]. When the continents are nearer the poles, as they are today, carbon dioxide in the atmosphere remains in high enough concentrations to keep the planet warm. When global temperatures drop enough that glaciers cover the high-latitude continents, as they do in Antarctica and Greenland, the ice sheets prevent chemical erosion of the rocks beneath the ice. With the carbon burial process stifled, the carbon dioxide in the atmosphere stabilizes at a level high enough to fend off the advancing ice sheets. If all the continents cluster in the tropics, on the other hand, they would remain ice-free even as the earth grew colder and approached the critical threshold for a runaway freeze. The carbon dioxide "safety switch" would fail because carbon burial continues unchecked.

We may never know the true trigger for a snowball earth, as we have but simple theories for the ultimate forcing of climate change, even in recent times. But we should be wary of the planet's capacity for extreme change. For the past million years, the earth has been in its coldest state since animals first appeared, but even the greatest advance of glaciers 20,000 years ago was far from the critical threshold needed to plunge the earth into a snowball state. Certainly during the next several hundred years, we will be more concerned with humanity's effects on climate as the earth heats up in response to carbon dioxide emissions [see "The Human Impact on Climate Change," by Thomas R. Karl and Kevin E. Trenberth; SCIENTIFIC AMERICAN, December 1999]. But could a frozen world be in our more distant future?

We are still some 80,000 years from the peak of the next ice age, so our first chance for an answer is far in the future. It is difficult to say where the earth's climate will drift over millions of years. If the trend of the past million years continues and if the polar continental safety switch were to fail, we may once again experience a global ice catastrophe that would inevitably jolt life in some new direction.

The Authors

Paul F. Hoffman and Daniel P. Schrag, both at Harvard University, bring complementary expertise to bear on the snowball earth hypothesis. Hoffman is a field geologist who has long studied ancient rocks to unravel the earth's early history. He led the series of expeditions to northwestern Namibia that turned up evidence for Neoproterozoic snowball earth events. Schrag is a geochemical oceanographer who uses the chemical and isotopic variations of coral reefs, deep-sea sediments and carbonate rocks to study climate on timescales ranging from months to millions of years. Together they were able to interpret the geologic and geochemical evidence from Namibia and to explore the implications of a snowball earth and its aftermath.

2 The Role of Vegetation and Methane in Climate Change

At long last, an article that suggests people are off the hook in terms of global warming, climate change, deforestation, and their far-reaching effects, such as desertification? Not entirely, but at least humans may not be to blame for the creation of the Sahara.

About 5,500 years ago, hippopotamuses wallowed in great lakes in the Sahara, surrounded by lush vegetation. Scientists once thought that when people began to arrive in the region, they overexploited the land (through hunting, farming, and clearing), triggering ecosystem collapse. But two computer simulations suggest that the area's vegetation patterns could have shifted quite quickly without people's help, due only to the slow climate change under way at the time.

Though humans may not have created the desert in northern Africa, the lesson it provides of radical ecosystem alteration in the midst of slow climate change should be troubling to people: Today, by slowly altering our climate through such human-driven mechanisms as global warming, deforestation, and erosion, we

*may be pushing ecosystems toward invisible
thresholds of irreversible change, with unfore-
seeable, and quite possibly destructive,
consequences. —KH*

"Deserting the Sahara"
by Sarah Simpson
Scientific American, October 1999

Plants may seem to sit passively as climate decides
their fate, but scientists are beginning to believe that
vegetation can strongly amplify the climate's most
subtle whims—sometimes with abrupt and devastating
results. A new computer simulation indicates that
plants helped to turn the Sahara from a lush grassland
thriving with hippos and elephants to its current
condition as the world's largest desert.

The Sahara's succulent sojourn faced an abrupt end
about 5,500 years ago. In a matter of centuries, rainfall
levels plummeted, the green grasslands paled to a sandy
yellow, and civilizations were forced to relocate. Many
scientists have assumed that human beings, who arrived
there 7,000 years ago, overused the land, which led to
the quick loss in vegetation. But the new simulations
show that a steady but slow loss of grasses—stemming
from a gradual trend toward less rainfall beginning
about 9,000 years ago—ran wildly out of control.

"Climate modelers tend to think that vegetation is
not important, because it's only about 20 percent of
the planet's surface area," says Martin Claussen, leader

of the team that designed the simulation at the Potsdam Institute for Climate Impact Research in Germany. "We're now seeing that we're not allowed to neglect land area."

John E. Kutzbach, a climatologist at the University of Wisconsin–Madison, is enthusiastic about the results because of their value in predicting future climate. "The idea that vegetation affects climate hasn't been studied in detail," he says.

Both Kutzbach and Claussen had independently used earlier computer simulations to watch how local weather affected Saharan plants, but Claussen's latest simulations allowed his team to be the first to see whether the plants themselves might effect change. Each of the team's 10 simulations began with the grasslands of 9,000 years ago and ended with the arid desert of the present.

The only external force they introduced to their simulated climate was a gradual evolution of the planet's orbit. About 9,000 years ago the earth's perihelion, the point at which the planet passes closest to the sun, occurred in July, and the North Pole was leaning more toward the sun. These two circumstances, which then meant stronger summer sunlight for the Northern Hemisphere and thus stronger monsoons to water a thirsty Saharan grassland, have changed slowly ever since. The northward tilt has shifted away from the sun, and perihelion now occurs in January.

During the first few thousand years of Claussen's simulation, this transition manifested as a gradual loss of vegetation, presumably because the monsoons were weakening. But the grassland's condition took a dramatic nosedive starting about 5,500 years ago, the same time that lakes and large animals begin to disappear from the fossil record. The team speculates that the grasses of the early Sahara trapped moisture that could evaporate and become new clouds—and new rain. As desert sands took over, less water recycled to the atmosphere, so even less rain fell and more plants died. "We can now explain the most important changes in Saharan climate without taking human beings into account," says team member Claudia Kubatzki.

Kutzbach says that the findings of Claussen's group "open up a research area rather than being the final word," but he agrees with their theory that a vicious feedback cycle between vegetation and the atmosphere could force dramatic changes. More specifically, the role that plants play in their own sustenance can be key to their destruction.

Just because the Sahara apparently dried up because of natural causes does not mean that humans are off the hook. Noting that as much as 30 percent of the rainfall in a tropical rain forest has cycled through the leaves and roots of its flora, Kutzbach and his colleagues suggest, based on their own climate simulations, that cutting down trees could produce a feedback cycle similar to what the earth's changing orbit set off in the Sahara. "If you deforest, the rain washes down the Amazon rather than going back into the clouds to form rain," Kutzbach says. And pumping ever more carbon dioxide and other greenhouse gases into the atmosphere may do more than slowly warm the planet.

"We can't specify what will happen," Claussen says, "but we're assessing the danger of climate surprises."

Most biology students learn how layered fossils on land preserve a chronological history of ancient environments, with deeper layers preserving evidence of organisms that are generally older than those found closer to the surface. Early paleontologists used those layers to piece together an initial understanding of the history of life on earth.

It turns out that similar layers of biological history can be found on ocean floors, preserved by tiny organisms called foraminifera (or forams). Forams need oxygen-rich water to survive, with different species thriving in different water temperatures. Their tiny, hard shells can preserve information about the ocean temperatures that predominated in their lifetimes. At death, forams drift down to the ocean floor, forming layer after layer of shells.

In this article, scientists describe how forams helped them understand a cataclysmic event that occurred in the earth's oceans 55 million years ago, when the seas apparently blasted methane gas into the air, changing the temperature of both land and sea. —KH

"Methane Fever"
by **Sarah Simpson**
Scientific American, **February 2000**

Not often does a past geologic event exemplify what the actions of humanity may inflict on the world. Most global changes, such as the waxing and waning of ice ages, take so long that they are indiscernible in human lifetimes. But 55 million years ago a series of methane gas blasts may have choked the atmosphere with greenhouse gases at a pace similar to that at which the burning of fossil fuels pumps them into the air today.

Back then, at the end of an epoch of time known as the Paleocene, temperatures in the deep ocean soared by about six degrees Celsius. This worldwide heat wave killed off a plethora of microscopic deep-sea creatures and produced a bizarre spike in the record of carbon isotopes. Five years ago paleoceanographer Gerald ("Jerry") Dickens of James Cook University in Australia proposed that a belch of seafloor methane—a greenhouse gas with almost 30 times the heat-trapping ability of carbon dioxide caused the shock. But no one had actually seen evidence of where this catastrophe might have happened until now.

Dickens, working with Miriam E. Katz of Rutgers University and two other researchers, recently discovered evidence of the exact sequence of predicted methane warming events buried under half a kilometer of sediment off Florida's northeastern coast. "It's the first really tangible evidence of methane release from that time," says marine geologist Timothy J. Bralower of the University of North Carolina at Chapel Hill. "It's almost too good to be true."

Katz, who helped to retrieve the prized seafloor sediment in 1997, was searching initially for the extinction. Some bottom-dwelling creatures called foraminifera, or forams, suffocated in the warmer water because it contains less oxygen than does cold water. Their hard shells were eventually buried in the seafloor muck.

Staring through a microscope for hours at a time, Katz painstakingly separated thousands of salt-grain-size forams from their muddy mass grave using a tiny paintbrush. Her search revealed that 55 percent of the species of deep-sea forams had disappeared from the fossil record in a blink of an eye in geologic time—less than 10,000 years within the late Paleocene climate fever. Katz's colleague Dorothy K. Pak of the University of California at Santa Barbara found that the shells of the surviving forams clearly recorded the carbon isotope spike.

Within the foram deathbeds, Katz was startled to notice a 25-centimeter thick layer of jumbled chunks of mud. "At first I complained that it was messing up my extinction event," Katz says. Then she remembered Dickens's idea about what might have caused the creatures to die in the first place: An explosion of methane escapes from seafloor hydrate deposits where the gas, generated as bacteria digest dead plants and animals, lies entombed in crystalline cages of ice. The gas then bubbles to the ocean surface, enters the atmosphere and begins trapping the heat that eventually warms the ocean water and suffocates the forams.

Such an explosion would have likely triggered a seafloor landslide, and the jumbled mud layer looked like the smoking gun of just such an event. That's when Katz called Dickens into the project. He based his original methane escape scenario on the fact that methane hydrate deposits, which today contain something like 15 trillion tons of gas, are the only place where organic methane exists in abundances that could alter the isotopic signature of the foram shells. When Dickens and Katz searched for the landslide source, they found chaotic sediment layers just downhill from a buried coral reef—an ideal place for gas bubbles to have gathered before freezing into icy hydrates.

Still, not everything is solved. Richard D. Norris of the Woods Hole Oceanographic Institution notes that an abrupt change in deep-ocean currents, rather than exploding hydrates, could explain the landslide. And what caused the methane to come out in the first place is not clear. One possible trigger is the five-million-year warming trend that led up to the end of the Paleocene and had already poised the planet for dramatic change. When the bottom waters reached a critical temperature, the fragile hydrates may have decomposed in a sudden blast.

Even so, Katz says, it would have taken a series of such blasts to generate the nearly one trillion tons of gas that Dickens calculated would have been necessary to account for the isotope spike. But besides melting, hydrates have another, shorter way of going from the seafloor to the sky. On a research cruise off the coast of Oregon last summer, Erwin Suess of the Research Center for Marine Geosciences in Kiel, Germany, and his colleagues saw refrigerator-size chunks of buoyant methane hydrate that had made a kilometer-long trip from the seafloor to the ocean surface before disintegrating.

A final question burns in Dickens's mind: "Once we get all of that carbon into the system, how do we get it out?" Understanding the consequences of the late Paleocene warming is crucial for the earth's current inhabitants. Even if we stopped driving our cars and burning coal in power plants today, Dickens says, the carbon dioxide that is already there would still have an impact down the line.

The following article provides a terrific example of a scientific mystery that remains unsolved to this day. What made the earth so warm before the first global ice age 2.3 billion years ago? The author, James F. Kasting, lists some of the possible chemical suspects. Carbon dioxide is an obvious one, given how involved it is in today's global warm-up. But if carbon dioxide levels were high before the first ice age, levels of certain types of minerals should have been high also. However, there is no evidence to suggest that they were. The plot thickens. What about ammonia? It's a powerful greenhouse gas, but it breaks down quickly in sunlight, so it wouldn't last long enough in the atmosphere to warm the earth dramatically.

After narrowing the various global warming suspects in this way, Kasting argues that methane could have done the job. He supports his case with both the gas's chemical composition and its source—microorganisms called methanogens. Kasting's thesis concerning the mechanisms and processes behind prehistoric global warming involves some key ideas that recur throughout the contemporary climate-change literature, such as feedback cycles and the link between the atmosphere and organisms. —KH

"When Methane Made Climate"
by James F. Kasting
Scientific American, July 2004

About 2.3 billion years ago unusual microbes breathed new life into young Planet Earth by filling its skies with oxygen. Without those prolific organisms, called cyanobacteria, most of the life that we see around us would never have evolved.

Now many scientists think another group of single-celled microbes were making the planet habitable long before that time. In this view, oxygen-detesting methanogens reigned supreme during the first two billion years of Earth's history, and the greenhouse effect of the methane they produced had profound consequences for climate.

Scientists first began to suspect methane's dramatic role more than 20 years ago, but only during the past four years have the various pieces of the ancient methane story come together. Computer simulations now reveal that the gas—which survives about 10 years in today's atmosphere—could have endured for as long as 10,000 years in an oxygen-free world. No fossil remains exist from that time, but many micro-biologists believe that methanogens were some of the first life-forms to evolve. In their prime, these microbes could have generated methane in quantities large enough to stave off a global deep freeze. The sun was considerably dimmer then, so the added greenhouse influence of methane could have been exactly what the planet needed to keep warm. But the methanogens did not dominate forever. The plummeting temperatures associated with their fading glory could explain Earth's first global ice age and perhaps others as well.

The prevalence of methane also means that a pinkish-orange haze may have shrouded the planet, as it does Saturn's largest moon, Titan. Although Titan's methane almost certainly comes from a nonbiological source, that moon's similarities to the early Earth could help reveal how greenhouse gases regulated climate in our planet's distant past.

Faint Sun Foiled

When Earth formed some 4.6 billion years ago, the sun burned only 70 percent as brightly as it does today [see "How Climate Evolved on the Terrestrial Planets," by James F. Kasting, Owen B. Toon and James B. Pollack; SCIENTIFIC AMERICAN, February 1988]. Yet the geologic record contains no convincing evidence for widespread glaciation until about 2.3 billion years ago, which means that the planet was probably even warmer than during the modern cycle of ice ages of the past 100,000 years. Thus, not only did greenhouse gases have to make up for a fainter sun, they also had to maintain average temperatures considerably higher than today's.

Methane was far from scientists' first choice as an explanation of how the young Earth avoided a deep freeze. Because ammonia is a much stronger greenhouse gas than methane, Carl Sagan and George H. Mullen of Cornell University suggested in the early 1970s that it was the culprit. But later research showed that the sun's ultraviolet rays rapidly destroy ammonia in an oxygen-free atmosphere. So this explanation did not work.

Another obvious candidate was carbon dioxide (CO_2), one of the primary gases spewing from the volcanoes abundant at that time. Although they debated the details, most scientists assumed for more than 20 years that this gas played the dominant role. In 1995, however, Harvard University researchers uncovered evidence that convinced many people that CO_2 levels were too low to have kept the early Earth warm.

The Harvard team, led by Rob Rye, knew from previous studies that if the atmospheric concentrations of CO_2 had exceeded about eight times the present-day value of around 380 parts per million (ppm), the mineral siderite ($FeCO_3$) would have formed in the top layers of the soil as iron reacted with CO_2 in the oxygen-free air. But when the investigators studied samples of ancient soils from between 2.8 billion and 2.2 billion years ago, they found no trace of siderite. Its absence implied that the CO_2 concentration must have been far less than would have been needed to keep the planet's surface from freezing.

Even before CO_2 lost top billing as the key greenhouse gas, researchers had begun to explore an alternative explanation. By the late 1980s, scientists had learned that methane traps more heat than an equivalent concentration of CO_2 because it absorbs a wider range of wavelengths of Earth's outgoing radiation. But those early studies underestimated methane's influence. My group at Pennsylvania State University turned to methane because we knew that it would have had a much longer lifetime in the ancient atmosphere.

In today's oxygen-rich atmosphere, the carbon in methane is much happier teaming up with the oxygen in hydroxyl radicals to produce CO_2 and carbon monoxide (CO), releasing water vapor in the process. Consequently, methane remains in the atmosphere a mere 10 years and plays just a bit part in warming the

Methane Makers on the Tree of Life

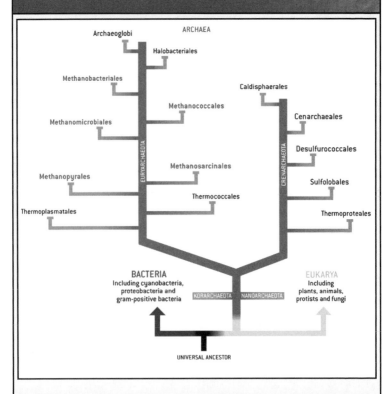

Methane-producing microbes called methanogens make up nearly half of all known Archaea, one of the three domains of living things—including Bacteria and Eukarya—that arose separately from an unknown ancestor. Methanogens exist in a variety of shapes, including rods and spheres, and live exclusively in oxygen-free settings. Because the oldest of the five orders of methanogens occupy low-lying branches of the Archaea domain, most biologists think these microbes were among the first organisms to evolve.

—*J. F. K.*

planet. Indeed, the gas exists in minuscule concentrations of only about 1.7 ppm; CO_2 is roughly 220 times as concentrated at the planet's surface and water vapor 6,000 times.

To determine how much higher those methane concentrations must have been to warm the early Earth, my students and I collaborated with researchers from the NASA Ames Research Center to simulate the ancient climate. When we assumed that the sun was 80 percent as bright as today, which is the value expected 2.8 billion years ago, an atmosphere with no methane at all would have had to contain a whopping 20,000 ppm of CO_2 to keep the surface temperature above freezing. That concentration is 50 times as high as modern values and seven times as high as the upper limit on CO_2 that the studies of ancient soils revealed. When the simulations calculated CO_2 at its maximum possible value, the atmosphere required the help of 1,000 ppm of methane to keep the mean surface temperature above freezing—in other words, 0.1 percent of the atmosphere needed to be methane.

Up to the Task?

The early atmosphere could have maintained such high concentrations only if methane was being produced at rates comparable to today. Were primordial methanogens up to the task? My colleagues and I teamed up with microbiologist Janet L. Siefert of Rice University to try to find out.

Biologists have several reasons to suspect that such high methane levels could have been maintained.

Siefert and others think that methane-producing microbes were some of the first microorganisms to evolve. They also suggest that methanogens would have filled niches that oxygen producers and sulfate reducers now occupy, giving them a much more prominent biological and climatic role than they have in the modern world.

Methanogens would have thrived in an environment fueled by volcanic eruptions. Many methanogens feed directly on hydrogen gas (H_2) and CO_2 and belch methane as a waste product; others consume acetate and other compounds that form as organic matter decays in the absence of oxygen. That is why today's methanogens can live only in oxygen-free environments such as the stomachs of cows and the mud under flooded rice paddies. On the early Earth, however, the entire atmosphere was devoid of oxygen, and volcanoes released significant amounts of H_2. With no oxygen available to form water, H_2 probably accumulated in the atmosphere and oceans in concentrations high enough for methanogens to use.

Based on these and other considerations, some scientists have proposed that methanogens living on geologically derived hydrogen might form the base of underground microbial ecosystems on Mars and on Jupiter's ice-covered moon, Europa. Indeed, a recent report from the European Space Agency's Mars Express spacecraft suggests that the present Martian atmosphere may contain approximately 10 parts per billion of methane. If verified, this finding would be

How Haze Forms

A methane-induced haze of hydrocarbon particles may have held the ancient Earth in a delicate balance between a hothouse and a deep freeze. The concentration of methane would have increased (*a*)—thereby intensifying the greenhouse effect (*b*)—for no more than a few tens of thousands of years before the climate-cooling haze would have developed (*c*).

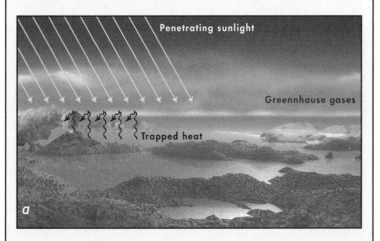

Penetrating sunlight

Greenhouse gases

Trapped heat

a

Methane's starring role in Earth's atmosphere may have begun almost as soon as life originated more than 3.5 billion years ago. Single-celled ocean dwellers called methanogens would have thrived in a world virtually devoid of oxygen—as Earth was at that time—and the methane they produced would have survived in the atmosphere much longer than it does today. This methane—along with another, more abundant greenhouse

continued on following page

continued from previous page

gas, carbon dioxide (CO_2) from volcanoes—would have warmed the planet's surface by trapping Earth's outgoing heat (*dark gray arrows*) while allowing sunlight (*light gray arrows*) to pass through.

A humid greenhouse is the preferred climate for many methanogens; the warmer the world became, the more methane they would have produced. This positive feedback loop would have strengthened the greenhouse effect, pushing surface temperatures even higher. Warmer conditions would have intensified the water cycle and enhanced the weathering of rocks on the continents—a process that pulls CO_2 out of the atmosphere. Concentrations of CO_2 would have dropped as those of methane continued to rise, until the two gases existed in nearly equal amounts. Under such conditions, methane would have altered dramatically.

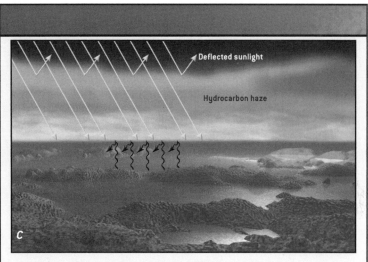

Deflected sunlight

Hydrocarbon haze

c

Changing chemistry would have kept the rising methane levels from turning Earth into a hothouse. Some of the methane would have linked together to form complex hydrocarbons that then condensed into dustlike particles. A high-altitude haze of these particles would have offset the intense greenhouse effect by absorbing the visible wavelengths of incoming sunlight and reradiating them back to space, thereby reducing the total amount of warmth that reached the planet's surface. Fewer heat-loving methanogens could have survived in the cooler climate; the haze would thus have put a cap on overall methane production.

consistent with having methanogens living below the surface of Mars.

Geochemists estimate that on the early Earth H_2 reached concentrations of hundreds to thousands of

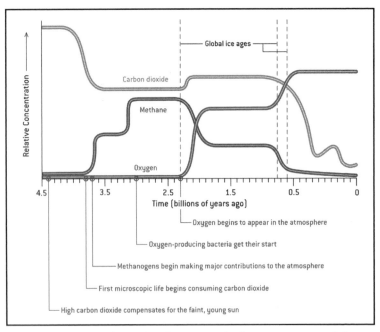

Relative concentrations of major atmospheric gases may explain why global ice ages (*dashed lines*) occurred in Earth's distant past. Methane-producing microorganisms flourished initially, but as oxygen skyrocketed about 2.3 billion years ago, these microbes suddenly found few environments where they could survive. The accompanying decrease in methane—a potent greenhouse gas—could have chilled the entire planet. The role of carbon dioxide, the most notable greenhouse gas in today's atmosphere, was probably much less dramatic.

parts per million—that is, until methanogens evolved and converted most of it to methane. Thermodynamic calculations reveal that if other essential nutrients, such as phosphorus and nitrogen, were available, methanogens would have used most of the available H_2 to make methane. (Most scientists agree that sufficient phosphorus would have come from the chemical breakdown of rocks and that various ocean-dwelling microorganisms were producing plenty of nitrogen.) In this scenario, the methanogens would have yielded the roughly 1,000 ppm of methane called for by the computer models to keep the planet warm.

Even more evidence for the primordial dominance of methanogens surfaced when microbiologists considered how today's methanogens would have reacted to a steamy climate. Most methanogens grow best at temperatures above 40 degrees Celsius; some even prefer at least 85 degrees C. Those that thrive at higher temperatures grow faster, so as the intensifying greenhouse effect raised temperatures at the planet's surface, more of these faster-growing, heat-loving specialists would have survived. As they made up a larger proportion of the methanogen population, more methane molecules would have accumulated in the atmosphere, making the surface temperature still warmer—in fact, hotter than today's climate, despite the dimmer sun.

Smog Saves the Day

As a result of that positive feedback loop, the world could have eventually become such a hothouse that life itself would have been difficult for all but the most extreme heat-loving microbes. This upward spiral could not have continued indefinitely, however. Once atmospheric methane becomes more abundant than CO_2, methane's reaction to sunlight changes. Instead of being oxidized to CO or CO_2, it polymerizes, or links together, to form complex hydrocarbons that then condense into particles, forming an organic haze. Planetary scientists observe a similar haze in the atmosphere of Saturn's largest moon: Titan's atmosphere consists primarily of molecular nitrogen, N_2, along with a small percentage of methane. The scientists hope to learn more when NASA's Cassini spacecraft arrives at Saturn in July [see "Saturn at Last!" by Jonathan I. Lunine; SCIENTIFIC AMERICAN, June 2004].

The possible formation of organic haze in Earth's young atmosphere adds a new wrinkle to the climate story. Because they form at high altitudes, these particles have the opposite effect on climate that greenhouse gases do. A greenhouse gas allows most visible solar radiation to pass through, but it absorbs and reradiates outgoing infrared radiation, thereby warming the surface. In contrast, high-altitude organic haze absorbs incoming sunlight and reradiates it back into space, thereby reducing the total amount of radiation that reaches the surface. On Titan, this so-called anti-greenhouse effect cools the surface by seven degrees C or so. A similar haze layer on the ancient Earth would have also cooled the climate, thus shifting the methanogen population back toward those slower-growing species that prefer cooler weather and thereby limiting further increases in methane production. This powerful negative feedback loop would have tended to stabilize Earth's temperature and atmospheric composition at exactly the point at which the layer of organic haze began to form.

Nothing Lasts Forever

Methane-induced smog kept the young Earth comfortably warm—but not forever. Global ice ages occurred at least three times in the period known as the Proterozoic eon, first at 2.3 billion years ago and again at 750 million and 600 million years ago. The circumstances surrounding these glaciations were long unexplained, but the methane hypothesis provides compelling answers here as well.

The first of these glacial periods is often called the Huronian glaciation because it is well exposed in rocks just north of Lake Huron in southern Canada. Like the better-studied late Proterozoic glaciations, the Huronian event appears to have been global, based on interpretations that some of the continents were near the equator at the time ice covered them.

This cold snap formed layers of jumbled rocks and other materials that a glacier carried and then dropped to the ground when the ice melted sometime between 2.45 billion and 2.2 billion years ago. In the older rocks below these glacial deposits are detrital uraninite and pyrite, two minerals considered evidence for very low levels of atmospheric oxygen. Above the glacial layers sits a red sandstone containing hematite—a mineral that forms only under oxygen-rich skies. (Hematite has also been found at the landing site of the Mars rover Opportunity. This hematite is gray, however, because the grain size is larger.) The layering of these distinct rock types indicates that the Huronian glaciations occurred precisely when atmospheric oxygen levels first rose.

This apparent coincidence remained unexplained until recently: if we hypothesize that methane kept the ancient climate warm, then we can predict a global ice age at 2.3 billion years ago because it would have been a natural consequence of the rise of oxygen. Many of the methanogens and other anaerobic organisms that dominated the planet before the rise of oxygen would have either perished in this revolution or found themselves confined to increasingly restricted habitats.

Although this finale sounds as if it is the end of the methane story, that is not necessarily the case. Methane never again exerted a dominating effect on climate, but it could still have been an important influence at later times—during the late Proterozoic, for example, when some scientists suggest that the oceans froze over entirely during a series of so-called snowball Earth episodes [see "Snowball Earth," by Paul F. Hoffman and Daniel P. Schrag; SCIENTIFIC AMERICAN, January 2000].

Indeed, methane concentrations could have remained significantly higher than today's during much of the Proterozoic eon, which ended about 600 million years ago, if atmospheric oxygen had continued to be somewhat lower and the deep oceans were still anoxic and low in sulfate, a dissolved salt common in modern seawater. The rate at which methane escaped from the seas to the atmosphere could still have been up to 10 times as high as it is now, and the concentration of methane in the atmosphere could have been as high as 100 ppm. This scenario might explain why the Proterozoic remained ice-free for almost a billion and a half years despite the fact that the sun was still relatively dim. My colleagues and I have speculated that a second rise in atmospheric oxygen, or in dissolved sulfate, could conceivably have triggered the snowball Earth episodes as well—once again by decreasing the warming presence of methane.

Extraterrestrial Methane

As compelling as this story of methanogens once ruling the world may sound, scientists are forced to be content with no direct evidence to back it up. Finding a rock that contains bubbles of ancient atmosphere would provide absolute proof, but such a revelation is unlikely. The best we can say is that the hypothesis is consistent with several indirect pieces of evidence—most notably, the low atmospheric CO_2 levels inferred from ancient soils and the timing of the first planet-encompassing ice age.

Although we may never be able to verify this hypothesis on Earth, we may be able to test it indirectly by observing Earth-like planets orbiting other stars. Both NASA and the European Space Agency are designing large space-based telescopes to search for Earth-size planets orbiting some 120 nearby stars. If such planets exist, these missions—NASA'S Terrestrial Planet Finder and ESA's Darwin—should be able to scan their atmospheres for the presence of gases that would indicate the existence of life.

Oxygen at any appreciable abundance would almost certainly indicate biology comparable to that of modern Earth, provided that the planet was also endowed with the liquid water necessary for life. High levels of methane, too, would suggest some form of life. As far as we know, on planets with Earth-like surface temperatures only living organisms can produce methane at high levels. The latter discovery might be one of the best ways for scientists to gain a better understanding of what our own planet was like during the nascent stages of its history.

The Author

James F. Kasting studies the origin and evolution of planetary atmospheres, especially those of Earth and its nearest neighbors, Venus and Mars. Since earning his Ph.D. in atmospheric science at the University of Michigan at Ann Arbor in 1979, he has used theoretical computer models to investigate atmospheric chemistry and to calculate the greenhouse effect of different gases and particles in both the present day and the distant past. Recently Kasting has begun exploring the question of whether Earth-like planets might exist around other stars in our galaxy. He is working with several other scientists to develop the theoretical foundation for NASA's Terrestrial Planet Finder, a space-based telescope designed to locate planets around other stars and scan their atmospheres for signs of life.

Not all climate research has resulted in gloom and doom conclusions. As the science of global climate change has matured, some studies have provided limited good news and optimistic moments. In one such study, scientists suggested that increased carbon dioxide in the atmosphere would act as a fertilizer, since plants "inhale" the gas and use it to build tissues. That certainly boded well for agriculture. Another study argued that the regrowing of forests in the northeastern United States and other parts of the world might lead to the absorption of much of the excess carbon dioxide we are currently pouring into the atmosphere. To a limited extent, both theories have validity, but research projects like the one described in the following article make clear that there are limits to such simple and straightforward atmosphere-cleansing strategies (strategies that may involve more than a little wishful thinking). In this article, for example, scientists argue that warmer nighttime temperatures harm tropical trees, slowing their growth and actually causing the trees to release more carbon dioxide into the atmosphere than they absorb from it. —KH

"In the Heat of the Night"
by Tim Beardsley
Scientific American, October 1998

Researchers working in Costa Rica have discovered disturbing evidence that increasing temperatures have markedly slowed the growth of tropical trees over the past decade. The slowdown may explain calculations suggesting that tropical forests, which are usually considered to take up carbon dioxide, have actually added billions of tons of the greenhouse gas to the atmosphere each year during the 1990s, making them a huge net source, comparable in size to the combustion of fossil fuels. The trend could exacerbate global warming: as the mercury rises, tropical forests may dump yet more carbon dioxide into the atmosphere, causing still more warming.

In 1984 researchers Deborah A. Clark and David B. Clark of the University of Missouri, collaborating with Charles D. Keeling and Stephen C. Piper of the Scripps Institution of Oceanography in La Jolla, Calif., began measuring the growth rates of scores of adult tropical rain-forest trees at La Selva Biological Station in central Costa Rica. The sample includes six different tree species, with both fast- and slow-growing types represented. Using special measuring collars, the scientists obtain reliable data on aboveground growth each year. Deborah Clark presented the team's findings in August at a meeting of the Ecological Society of America in Baltimore.

The group found that growth of all the trees fluctuated considerably from year to year. Moreover, the year-to-year changes correspond strikingly with the results of separate calculations of the size each year of a colossal unexplained tropical terrestrial source of carbon dioxide. In years when this theoretical source was large, the trees had grown slowly; in years when it was small or negative, the trees had grown faster.

The apparent lesson is that the varying annual growth rate of trees in tropical forests could account, in large part, for a calculated increase in carbon dioxide released from land in the tropical zone in the 1980s and 1990s (although other sources, such as soil microbes, probably also contribute). Although trees take in carbon dioxide and release oxygen during photosynthesis, they also release some carbon dioxide as a by-product of respiration, as most organisms do. When growing vigorously, plants take up more than they produce. But if growth slows, the balance shifts.

The annual excess of carbon coming from tropical forests, according to a preliminary calculation by Keeling and his associates, has been more than four billion tons in some recent years. Many researchers regard such estimates as provocative but not ironclad. The new data on tree growth "increase confidence in Keeling's work," Clark says. For comparison, worldwide carbon release into the atmosphere from the combustion of fossil fuels is estimated to be about 6.5 billion tons each year.

In an effort to understand what was causing the year-to-year variations in the rate of tree growth in

Costa Rica, Clark and her colleagues evaluated climatic factors. They found that rate of growth was strongly linked to average temperature, slowing down in warmer years. The negative link was even stronger between growth rate and daily minimum temperature. "Tropical trees are being increasingly stressed through higher nighttime temperatures," Clark concludes. She thinks higher nighttime temperatures force the trees to respire more, thus promoting release of carbon dioxide. Yet warming does not increase photosynthesis, leading to a growing imbalance.

The new information from Costa Rica has not yet been published in a peer-reviewed journal, so it remains to be seen whether the scientific community will accept it. Globally, daily minimum temperatures have been increasing faster than average temperatures, so the data suggest that tropical forests might become an even bigger net source of carbon dioxide in coming years. On the other hand, studies of trees in temperate regions indicate that artificially increased levels of carbon dioxide cause trees there to grow faster, which in principle might counter the heat-induced suppression of tree growth. But Clark's observations seem to suggest that the growth-slowing effect of increased temperatures in tropical regions is now stronger than any beneficial fertilizing effect from rising carbon dioxide.

Lest anyone get the mistaken idea that destroying tropical forests would help, James T. Randerson of the Carnegie Institution of Washington notes that clearing a forest adds much more of the gas to the atmosphere than does leaving it be. Researchers believe that tropical forests account for about one third of all carbon dioxide taken out of the atmosphere by photosynthesis on land, making them a crucial part of the global atmospheric equation. The newly detected slowing effect of temperature on tropical forest growth "could be a positive feedback" that will speed global warming, Clark warns.

3 The Present and Near-Future Effects of Global Warming

The debate about the relationship between global warming and the increasing incidence and intensity of hurricanes, described briefly in the next article, has only become more and more rancorous, especially in the wake of the destruction wreaked along the Gulf Coast in late August and early September 2005 by Hurricanes Katrina and Rita. Some climate change scientists say the raging debate has come to illustrate the increasing politicization of their work. In 2005, scientists blasted each other in the media over competing claims about global warming and hurricane formation, and some of those named in this article entered the fray. Interestingly, Kerry A. Emanuel, a meteorologist quoted in the article, published a paper in a prestigious journal that same year reviewing about fifty years of hurricane and typhoon data. Emanuel concluded that the storms were lasting significantly longer and hitting with greater wind intensity, a conclusion seemingly later borne out by Katrina and Rita.

The final point here is an intriguing and important one to discuss: Regardless of the debate

about the validity of the science linking global
warming to an increase in Atlantic hurricanes,
it is clear that human population shifts are
contributing to the increasing destructiveness of
these storms. More and more people are moving
to vulnerable locations along the Atlantic and
Gulf coasts. They are building bigger and flimsier
houses and more expansive businesses, exposing
themselves to the fury of these storms. —KH

"Stormy Weather"
by Mark Alpert
Scientific American, December 2004

Florida residents will long remember the hurricane
season of 2004. From early August to late September,
six major hurricanes (category 3 or above, in which
maximum wind speeds hit at least 178 kilometers per
hour) formed in the North Atlantic basin. Four of
them—Charley, Frances, Ivan and Jeanne—slammed
into the Sunshine State. (Ivan's eye actually made
landfall in Alabama, but the hurricane's winds roughed
up Florida's panhandle.) Although the targeting of
Florida seems to be mostly a case of bad luck—the
tracks of Atlantic hurricanes depend on the chaotic
vagaries of pressure highs and lows along the eastern
seaboard—many researchers are convinced that overall
hurricane activity in the Atlantic is on the upswing.

Since 1995 the annual number of major Atlantic
hurricanes has averaged 3.8, significantly higher than

the 60-year average of 2.3. In fact, the occurrence of these hurricanes seems to be oscillating on a decades-long cycle, with activity high from the late 1920s to the 1960s, low from 1970 to 1994 and then rebounding about 10 years ago. The oscillation is by no means smooth; hurricane activity in the Atlantic also swings sharply from year to year. (Overall, however, global hurricane activity is remarkably stable—busy seasons in one ocean are typically counterbalanced by calm seasons in another.)

Many researchers believe the year-to-year changes may be partly the result of El Niño warming events in the eastern Pacific Ocean, which may disrupt the formation of Atlantic hurricanes by increasing the difference between wind speeds at upper and lower altitudes. La Niña cooling events may have the opposite effect. But the reasons for the longterm hurricane trends are more mysterious.

Some scientists have searched for corresponding trends in the thermodynamics of the Atlantic Ocean. Hurricanes can form only over waters that are warmer than 26.5 degrees Celsius, and sea-surface temperatures in the North Atlantic were relatively high during the decades of above-average hurricane activity and low during the inactive period. William M. Gray, a veteran hurricane researcher at Colorado State University, believes the long-term hurricane cycle may be linked to global ocean currents that bring warm salty water from the tropics to the far North Atlantic. When this thermohaline circulation is strong, the North Atlantic warms, and more major hurricanes are born; when the circulation weakens, perhaps because of an injection of freshwater from Arctic ice, hurricane activity decreases. According to Gray, the eastern U.S. will have to endure an above-average number of major hurricanes for the next 20 to 30 years. "I'll be in my grave before it's over," says Gray, who is 74.

Other researchers are skeptical of this neat picture because the process of hurricane creation is so devilishly complicated. Meteorologist Kerry A. Emanuel of the Massachusetts Institute of Technology notes that hurricane genesis depends not so much on ocean temperatures alone but on the difference in temperature between the sea surface and the upper atmosphere. James B. Eisner, a climatologist at Florida State University, is focusing his attention on the North Atlantic Oscillation (NAO), a poorly understood climate mode that periodically shifts the tracks of storms crossing the ocean. When the NAO weakens, a pressure high moves southwest from the Azores to Bermuda; this prevents hurricanes from turning north, so they gain strength and head for the Caribbean and the southeastern U.S. The NAO weakened dramatically in late July, just before the spate of hurricanes.

Whatever the cause of the renewed activity, scientists agree that Florida and other southeastern states are particularly vulnerable because so much seacoast development occurred during the 25-year lull in major hurricanes. Because a weak El Niño event is currently under way, the 2005 hurricane season may well be more moderate, but catastrophic storms could return in force in following years. Says Eisner: "There is some indication that 2004 is a harbinger of things to come."

Many scientists see the changes occurring in the Arctic—most important, the retreating of sea ice and melting of the polar cap—as signs of impending change in the earth's lower latitudes, where most of us live. The correlation between climate change in the Arctic and temperate regions is based on two premises. First, climate models have long predicted that global warming would occur first and fastest at the earth's poles. More important, as this article deftly points out, the Arctic serves as the earth's air conditioner in many ways. Changes in Arctic climate can ripple across the globe, through shifts in global ocean currents or winds.

Conditions are shifting fast in the Arctic, and people who live there are already seeing the effects. As sea ice retreats, winds can push up larger waves in the open ocean. Those waves, slamming against the shore, have eroded shore-lines, forcing whole villages in Alaska to move inland. Just a few years before this article was written, one of its authors, Mark C. Serreze, called himself a "fence sitter" in the debate about whether humans were causing global warming. He suspected a natural climate cycle might be the culprit. By 2005, however, Serreze said he had seen too many glaciers crumble into the ocean and too much sea ice retreat to remain on the fence. He became an unhappy convert to the belief in the human impact upon global warming and climate change. —KH

"Meltdown in the North"
by Matthew Sturm, Donald K. Perovich, and Mark C. Serreze
Scientific American, October 2003

Snow crystals sting my face and coat my beard and the ruff of my parka. As the wind rises, it becomes difficult to see my five companions through the blowing snow. We are 500 miles into a 750-mile snowmobile trip across Arctic Alaska. We have come, in the late winter of 2002, to measure the thickness of the snow cover and estimate its insulating capacity, an important factor in maintaining the thermal balance of the permafrost. I have called a momentary halt to decide what to do. The rising wind, combined with −30 degree Fahrenheit temperatures, makes it clear we need to find shelter, and fast. I put my face against the hood of my nearest companion and shout: "Make sure everyone stays close together. We have to get off this exposed ridge."

At the time, the irony that we might freeze to death while looking for evidence of global warming was lost on me, but later, snug in our tents, I began to laugh at how incongruous that would have been. —Matthew Sturm

The list is impressively long: The warmest air temperatures in four centuries, a shrinking sea-ice cover, a record amount of melting on the Greenland Ice Sheet, Alaskan glaciers retreating at unprecedented rates. Add to this the increasing discharge from Russian rivers, an Arctic growing season that has lengthened by several days per decade, and permafrost that has started to thaw. Taken together, these observations announce in a way no single measurement could that the Arctic is undergoing a profound transformation. Its full extent has come to light only in the past decade, after scientists in different disciplines began comparing their findings. Now many of those scientists are collaborating, trying to understand the ramifications of the changes and to predict what lies ahead for the Arctic and the rest of the globe.

What they learn will have planetwide importance because the Arctic exerts an outsize degree of control on the climate. Much as a spillway in a dam controls the level of a reservoir, the polar regions control the earth's heat balance. Because more solar energy is absorbed in the tropics than at the poles, winds and ocean currents constantly transport heat poleward, where the extensive snow and ice cover influences its fate. As long as this highly reflective cover is intact and extensive, sunlight coming directly into the Arctic is mostly reflected back into space, keeping the Arctic cool and a good repository for the heat brought in from lower latitudes. But if the cover begins to melt and shrink, it will reflect less sunlight, and the Arctic will become a poorer repository, eventually warming the climate of the entire planet.

Figuring out just what will happen, however, is fraught with complications. The greatest of these stems from the intricate feedback systems that govern the climate in the Arctic. Some of these processes are positive, amplifying change and turning a nudge into a shove, and some are negative, behaving as a brake on the system and mitigating change.

Chief among these processes is the ice-albedo feedback, in which rising temperatures produce shorter winters and less extensive snow and ice cover, with ripple effects all the way back through the midlatitudes. Another feedback is associated with the large stores of carbon frozen into the Arctic in the form of peat. As the climate warms and this peat thaws, it could release carbon dioxide into the atmosphere and enhance warming over not just the Arctic but the whole globe—a phenomenon commonly referred to as greenhouse warming.

The key problem is that we don't fully understand how some of these feedback processes work in isolation, let alone how they interact. What we do know is that the Arctic is a complex system: change one thing, and everything else responds, sometimes in a counter-intuitive way.

Heating Up

The more we look, the more change we see. Arctic air temperatures have increased by 0.5 degree Celsius each decade over the past 30 years, with most of the warming coming in winter and spring. Proxy records (ice and peat cores, lake sediments), which tell us mostly about summer temperatures, put this recent warming in perspective. They indicate that late 20th- and early 21st-century temperatures are at their highest level in 400 years. The same records tell us that these high levels are the result of steady warming for 100 years as the Arctic emerged from the Little Ice Age, a frigid period that ended around 1850, topped off by a dramatic acceleration of the warming in the past half a century.

The recent temperature trends are mirrored in many other time series. One example is that Arctic and Northern Hemisphere river and lake ice has been forming later and melting earlier since the Little Ice Age. The total ice-cover season is 16 days shorter than it was in 1850. Near one of our homes (Sturm's) in Alaska, a jackpot of about $300,000 awaits the person who can guess the date the Tanana River will break up every spring. The average winning date has gotten earlier by about six days since the betting pool was instituted in 1917. Higher-tech data—satellite images— show that the snow-free season in the Arctic has lengthened by several days each decade since the early 1970s. Similarly, the growing season has increased by as much as four days.

Shrinking Glaciers, Thawing Permafrost

There was nothing complex about my first research in Arctic climate change: march around a small glacier on Rilesmere Island, drill holes in the ice, insert long metal poles in the holes, measure them, come back a year later and see if more pole was showing.

We put in most of the pole network in the warm summer of 1982 and returned in 1983 to a very different world—week after week of cold, snow and fog. It looked like the start of a new ice age. Our plan had been to go back annually, but as so often happens, funding dried up, and my Arctic experiences became fond memories.

But memories sometimes get refreshed. In 2002 I got a call from an excited graduate student. He had revisited the glacier. It was rapidly wasting away. 1983 had been an anomaly. My stakes were there, except they were all lying on the surface of the ice. How deeply had I installed them? Did I still have my field notes? He need not have worried. There was my field book, dusty but safe in my bookcase. Now I'm going back to Rilesmere Island, to see what's left of the glacier that in 1983 seemed like such a permanent feature of the landscape but that I now realize may well die before I do. —Mark C. Serreze

Arctic glaciers tell a striking tale as well. In Alaska, they have been shrinking for five decades, and more startlingly, the rate of shrinkage has increased three-fold in the past 10 years. The melting glaciers translate into a rise in sea level of about two millimeters a

decade, or 10 percent of the total annual rise of
20 millimeters. Determining the state of the much
larger and more slowly changing Greenland Ice Sheet
has been something of a Holy Grail for Arctic
researchers. Older field and satellite results suggested
that the ice sheet was exhibiting asymmetrical
behavior—the west side thinning in a modest way
and the east side remaining in balance. Recent satellite
images indicate that the melt rate over the entire ice
sheet has been increasing with time. The total area
melting in a given summer has increased by 7 percent
each decade since 1978, with last summer setting an
all-time record. Winter snowfall appears insufficient
to offset this heavy summer melt, so the sheet is
shrinking.

The permafrost—the permanently frozen layer
below the surface—is thawing, too. In a study published
in 1986, researchers from the U.S. Geological Survey
carefully logged temperature profiles in deep oil-
exploration boreholes drilled through the permafrost
of northern Alaska. When they extrapolated the
profiles to the surface, they found an anomalous cur-
vature that was best explained by a warming at
ground level of two to four degrees C during the
preceding few decades. More recently, preliminary
results suggest an additional increase of two to three
degrees C has occurred since 1986. Because the
Arctic winter lasts nine months of the year, snow
cover controls the thermal state of the ground as
much as air temperature does, so these borehole

records almost certainly reflect a change in the amount and timing of winter precipitation as well as an increase in temperature. More snow means thicker insulation and therefore better protection for the ground from frigid winter temperatures. Ground that is not chilled as much in the winter is primed for more warming in the summer.

Regardless of why it is occurring, one thing is certain. Thawing permafrost is trouble. It can produce catastrophic failure of roads, houses and other infrastructure. It is also implicated in another recently detected change: over the past 60 years, the discharge of freshwater from Russian rivers into the Arctic Basin has increased by 7 percent—an amount equivalent to roughly one quarter the volume of Lake Erie or three months of the outflow of the Mississippi River. Scientists attribute the change partly to greater winter precipitation and partly to a warming of the permafrost and active layer, which they believe is now transporting more groundwater. This influx of freshwater could have important implications for global climate: the paleo-record suggests that when the outflow of water from the Arctic Basin hits a critical level of freshness, the global ocean circulation changes dramatically. When ocean circulation changes, climate does as well, because the circulation system—essentially a set of moving rivers of water in the ocean, such as the Gulf Stream—is one of the prime conveyors of heat northward toward the pole.

Greening of the Arctic

The arctic land cover is also shifting. Based on warming experiments using greenhouses, biologists have known for some time that shrubs will grow at the expense of the other tundra plants when the climate warms. Under the same favorable growing conditions, the tree line will migrate north. Researchers have been looking for these modifications in the real world, but ecosystem responses can be slow. Only in the past few years, by comparing modern photographs with ones taken 50 years ago, and by using satellites to detect the increasing amount of leaf area, have researchers been able to document that both types of transformations are under way. As the vegetation alters, so does the role of the Arctic in the global carbon cycle. Vast stores of carbon in the form of peat underlie much of the tundra in Alaska and Russia, evidence that for long periods Arctic tundra has been a net carbon sink; about 600 cubic miles of peat are currently in cold storage. In recent years, warming has produced a shift: the Arctic now appears to be a net source of carbon dioxide. The change is subtle but troubling because carbon dioxide and methane constitute the primary greenhouse gases in the atmosphere, returning heat to the earth instead of allowing it to escape into space.

Warmer winters have driven some of the shift. When the air is warmer, more precipitation falls from the sky, some of it coming as snow. The thicker snow holds more warmth in the earth, resulting in a longer period during which the tundra is releasing carbon dioxide. But as the tundra becomes shrubbier and as the soil becomes drier in the summer as a result of higher temperatures, the balance could swing back the other way, because plants, particularly woody ones, will fix more carbon and lock it back into the Arctic ecosystem. The most recent studies suggest, in fact, that the magnitude and direction of the Arctic carbon balance depend on the time span that we are examining, with the response varying as the plants adapt to the new conditions.

Melting Sea Ice

"This sea ice is ridiculously thin," I thought as I broke through the ice for the second time that morning in August 1998. There was no real danger, now that personal flotation devices had become the de rigueur fashion accessory, but the thin ice was troubling for other reasons.

My journey to this place, 600 miles from the North Pole, had begun 10 months earlier on board the icebreaker Des Groseilliers, *which we had intentionally frozen into the pack to begin a yearlong drift. Our mission was to study ice-albedo and cloud-radiation feedbacks. When we started the journey, I was surprised at how thin the ice was. Now, after a much longer than expected summer melt season, it was thinner still, even though we had been drifting steadily north. I was uncertain which would come first: the end of the summer or the end of the ice. Little did I know that this summer the record for minimum ice cover was being set throughout the entire western Arctic Ocean.*

Unfortunately for the long-term survival of the ice pack, it was a record that was easily broken in 2002. —Donald K. Perovich

Of all the changes we have catalogued, the most alarming by far has been the reduction in the Arctic sea-ice cover. Researchers tracking this alteration have discovered that the area covered by the ice has been decreasing by about 3 percent each decade since the advent of satellite records in 1972. This rate might be low for a financial investment, but where time is measured in centuries or millennia, it is high. With the sea ice covering an area approximately the size of the U.S., the reduction per decade is equivalent to an area the size of Colorado and New Hampshire combined, the home states of two of us (Perovich and Serreze). The change in the thickness of the ice (determined from submarines) is even more striking: as much as 40 percent lost in the past few decades. Some climate models suggest that by 2080 the Arctic Ocean will be ice-free in summer.

The melting sea ice does not raise sea level as melting glaciers do, because the ice is already floating, but it is alarming for two other reasons. Locally, the demise of the sea ice leads to the loss of a unique marine ecosystem replete with polar bears, seals and whales. Globally, an ice-free Arctic Ocean would be the extreme end point of the ice-albedo feedback—far more solar radiation would be absorbed, warming not just the Arctic but eventually every part of the earth.

The shrinking sea-ice cover has not escaped the attention of businesspeople, tourists and politicians. Serious discussions have been under way about the feasibility of transporting cargo via Arctic waters— including through the fabled Northwest Passage, now perhaps close to being a practical shipping route because of climate change. Roald Amundsen, the redoubtable Norwegian polar explorer, took more than three years to complete the first transit of the passage in 1906, when the Arctic was still under the influence of the Little Ice Age. Many explorers before him had died trying to make the journey. In the past few years, however, dozens of ships have completed the route, including Russian icebreakers refurbished for the tourist trade. These events would have been unimaginable, even with icebreakers, in the more intense ice conditions of 100 years ago.

Is Greenhouse Warming the Culprit?

This inventory of startling transformation in the Arctic inevitably raises the question of whether we are still emerging from the Little Ice Age or whether something quite different is now taking place. Specifically, should we interpret these changes as being caused by the increased concentration of atmospheric greenhouse gases overriding a natural temperature cycle? Or are they part of a longer-than-expected natural cycle?

The intricate web of feedback interactions renders this question exceedingly complicated—and we don't know enough yet to answer it unequivocally. But we know enough to be very worried.

Whatever is causing the melting and thawing now wracking the Arctic, these modifications have initiated a cascade of planetwide responses that will continue even if the climate were suddenly and unexpectedly to stop warming. Imagine the climate as a big, round rock perched on uneven terrain. The inventory tells us that the rock has been pushed a little—either by a natural climate cycle or by human activity—and has started to roll. Even if the pushing stops, the rock is going to keep rolling. When it finally does stop, it will be in a completely different place than before.

To cope with the constellation of changes in the Arctic in a concerted fashion and to develop an ability to predict what will happen next rather than just react to it, several federal agencies have begun to coordinate their Arctic research in a program called SEARCH (Study of Environmental Arctic Change). Early results give some promise for success in teasing out the linkages among the tightly coupled systems that shape the climate of the Arctic and thus the earth. A recent discovery about the patterns of wind circula-tion, for example, helps to explain previously puzzling spatial patterns of increasing temperature. Equally important, high-quality records of climate change now extend back 30 to 50 years.

Soon these records and other findings should allow us to determine whether the Arctic transformation is a natural trend linked to emergence from the Little Ice

Age or something more ominous. Our most difficult challenge in getting to that point is to come to grips with how the various feedbacks in the Arctic system interact—and to do so quickly.

The Authors

Matthew Sturm, Donald K. Perovich, and Mark C. Serreze have spent most of their research careers trying to understand the snow, ice and climate of the Arctic. In 16 years at the U.S. Army Cold Regions Research and Engineering Laboratory-Alaska, Sturm has led more than a dozen winter expeditions in Arctic Alaska, including most recently a 750-mile snowmobile traverse across the region. Perovich is with the New Hampshire office of the U.S. Army Cold Regions Research and Engineering Laboratory. His work has focused on sea ice and the ice-albedo feedback. Perovich was chief scientist on Ice Station SHEBA, a yearlong drift of an icebreaker frozen into the Arctic pack ice. Since 1986 Serreze has been with the National Snow and Ice Data Center at the University of Colorado at Boulder. His studies have emphasized Arctic climate change and interactions between sea ice and the atmosphere.

Paul Epstein sounds the alarm in this descriptive article about how climate change could affect human health. As if it is not distressing enough to contemplate rising ocean levels and longer-lasting droughts, Epstein predicts that a warmer world will also be an unhealthier one for people. His assessment, based on some of his own work and that of others, emerges logically from the starting point of climate change. Climate change is likely to increase heat and weather extremes, change ecosystems, alter the water cycle, and force people to shift their lifestyles and habits, including where they are able to live comfortably and safely. Any one of these climate-driven changes can lead to new or emerging diseases. People leaving a drought-stricken area, for example, might crowd into a new town, increasing the frequency of diseases related to population density, such as tuberculosis. Alternating periods of drought and rain can allow mosquitoes that carry malaria and dengue fever to thrive. Epstein acknowledges his thesis has its weaknesses. Outbreaks of disease can be triggered by so many things that it is often hard to pin the blame on any one cause. Even so, his article is both thought provoking and hair raising. —KH

"Is Global Warming Harmful to Health?"
by Paul R. Epstein
Scientific American, August 2000

Today few scientists doubt the atmosphere is warming. Most also agree that the rate of heating is accelerating and that the consequences of this temperature change could become increasingly disruptive. Even high school students can reel off some projected outcomes: the oceans will warm, and glaciers will melt, causing sea levels to rise and salt water to inundate settlements along many low-lying coasts. Meanwhile the regions suitable for farming will shift. Weather patterns should also become more erratic and storms more severe.

Yet less familiar effects could be equally detrimental. Notably, computer models predict that global warming, and other climate alterations it induces, will expand the incidence and distribution of many serious medical disorders. Disturbingly, these forecasts seem to be coming true.

Heating of the atmosphere can influence health through several routes. Most directly, it can generate more, stronger and hotter heat waves, which will become especially treacherous if the evenings fail to bring cooling relief. Unfortunately, a lack of nighttime cooling seems to be in the cards; the atmosphere is heating unevenly and is showing the biggest rises at night, in winter and at latitudes higher than about 50 degrees. In some places, the number of deaths related to heat waves is projected to double by 2020. Prolonged heat can, moreover, enhance production of smog and the dispersal of allergens. Both effects have been linked to respiratory symptoms.

Global warming can also threaten human well-being profoundly, if somewhat less directly, by revising weather patterns—particularly by pumping up the frequency and

intensity of floods and droughts and by causing rapid
swings in the weather. As the atmosphere has warmed
over the past century, droughts in arid areas have
persisted longer, and massive bursts of precipitation
have become more common. Aside from causing death
by drowning or starvation, these disasters promote by
various means the emergence, resurgence and spread
of infectious disease.

That prospect is deeply troubling, because infectious
illness is a genie that can be very hard to put back into
its bottle. It may kill fewer people in one fell swoop
than a raging flood or an extended drought, but once it
takes root in a community, it often defies eradication
and can invade other areas.

The control issue looms largest in the developing
world, where resources for prevention and treatment
can be scarce. But the technologically advanced
nations, too, can fall victim to surprise attacks—as
happened last year when the West Nile virus broke out
for the first time in North America, killing seven New
Yorkers. In these days of international commerce and
travel, an infectious disorder that appears in one part
of the world can quickly become a problem continents
away if the disease-causing agent, or pathogen, finds
itself in a hospitable environment.

Floods and droughts associated with global climate
change could undermine health in other ways as well.
They could damage crops and make them vulnerable
to infection and infestations by pests and choking
weeds, thereby reducing food supplies and potentially

contributing to malnutrition. And they could permanently or semipermanently displace entire populations in developing countries, leading to over-crowding and the diseases connected with it, such as tuberculosis.

Weather becomes more extreme and variable with atmospheric heating in part because the warming accelerates the water cycle: the process in which water vapor, mainly from the oceans, rises into the atmosphere before condensing out as precipitation. A warmed atmosphere heats the oceans (leading to faster evaporation), and it holds more moisture than a cool one. When the extra water condenses, it more frequently drops from the sky as larger downpours. While the oceans are being heated, so is the land, which can become highly parched in dry areas. Parching enlarges the pressure gradients that cause winds to develop, leading to turbulent winds, tornadoes and other powerful storms. In addition, the altered pressure and temperature gradients that accompany global warming can shift the distribution of when and where storms, floods and droughts occur.

I will address the worrisome health effects of global warming and disrupted climate patterns in greater detail, but I should note that the consequences may not all be bad. Very high temperatures in hot regions may reduce snail populations, which have a role in transmitting schistosomiasis, a parasitic disease. High winds may at times disperse pollution. Hotter winters in normally chilly areas may reduce cold-related heart attacks and respiratory ailments. Yet overall, the undesirable effects of more variable weather are likely to include new stresses and nasty surprises that will overshadow any benefits.

Mosquitoes Rule in the Heat

Diseases relayed by mosquitoes—such as malaria, dengue fever, yellow fever and several kinds of encephalitis—are among those eliciting the greatest concern as the world warms. Mosquitoes acquire disease-causing microorganisms when they take a blood meal from an infected animal or person. Then the pathogen reproduces inside the insects, which may deliver disease-causing doses to the next individuals they bite.

Mosquito-borne disorders are projected to become increasingly prevalent because their insect carriers, or "vectors," are very sensitive to meteorological conditions. Cold can be a friend to humans, because it limits mosquitoes to seasons and regions where temperatures stay above certain minimums. Winter freezing kills many eggs, larvae and adults outright. *Anopheles* mosquitoes, which transmit malaria parasites (such as *Plasmodium falciparum*), cause sustained outbreaks of malaria only where temperatures routinely exceed 60 degrees Fahrenheit. Similarly, *Aedes aegypti* mosquitoes, responsible for yellow fever and dengue fever, convey virus only where temperatures rarely fall below 50 degrees F.

Excessive heat kills insects as effectively as cold does. Nevertheless, within their survivable range of

Risk of malaria transmission will have risen in many parts of the world by 2020 (relative to the average risk in the years 1961 to 1990), according to projections assuming a temperature increase of about two degrees Fahrenheit. The analysis was based solely on temperature threshold and did not assess other factors that could influence malaria's spread.

temperatures, mosquitoes proliferate faster and bite more as the air becomes warmer. At the same time, greater heat speeds the rate at which pathogens inside them reproduce and mature. At 68 degrees F, the immature *P. falciparum* parasite takes 26 days to develop fully, but at 77 degrees F, it takes only 13 days. The *Anopheles* mosquitoes that spread this malaria parasite live only several weeks; warmer temperatures raise the odds that the parasites will mature in time for the mosquitoes to transfer the infection. As whole areas heat up, then, mosquitoes could expand into formerly forbidden territories, bringing illness with them. Further, warmer nighttime and winter temperatures may enable them to cause more disease for longer periods in the areas they already inhabit.

The extra heat is not alone in encouraging a rise in mosquito-borne infections. Intensifying floods and droughts resulting from global warming can each help trigger outbreaks by creating breeding grounds for insects whose desiccated eggs remain viable and hatch in still water. As floods recede, they leave puddles. In times of drought, streams can become stagnant pools, and people may put out containers to catch water; these pools and pots, too, can become incubators for new mosquitoes. And the insects can gain another boost if climate change or other processes (such as alterations of habitats by humans) reduce the populations of predators that normally keep mosquitoes in check.

Mosquitoes on the March

Malaria and dengue fever are two of the mosquito-borne diseases most likely to spread dramatically as global temperatures head upward. Malaria (marked by chills, fever, aches and anemia) already kills 3,000 people, mostly children, every day. Some models project that by the end of the 21st century, ongoing warming will have enlarged the zone of potential malaria transmission from an area containing 45 percent of the world's population to an area containing about 60 percent. That news is bad indeed, considering that no vaccine is available and that the causative parasites are becoming resistant to standard drugs.

True to the models, malaria is reappearing north and south of the tropics. The U.S. has long been home to *Anopheles* mosquitoes, and malaria circulated here decades ago. By the 1980s mosquito-control programs and other public health measures had restricted the disorder to California. Since 1990, however, when the hottest decade on record began, outbreaks of locally transmitted malaria have occurred during hot spells in Texas, Florida, Georgia, Michigan, New Jersey and New York (as well as in Toronto). These episodes undoubtedly started with a traveler or stowaway mosquito carrying malaria parasites. But the parasites clearly found friendly conditions in the U.S.—enough warmth and humidity, and plenty of mosquitoes able to transport them to victims who had not traveled. Malaria has returned to the Korean peninsula, parts of southern Europe and the former Soviet Union and to the coast of South Africa along the Indian Ocean.

Dengue, or "breakbone," fever (a severe flu-like viral illness that sometimes causes fatal internal bleeding) is spreading as well. Today it afflicts an estimated 50 million to 100 million in the tropics and subtropics (mainly in urban areas and their surroundings). It has broadened its range in the Americas over the past 10 years and had reached down to Buenos Aires by the end of the 1990s. It has also found its way to northern Australia. Neither a vaccine nor a specific drug treatment is yet available.

Although these expansions of malaria and dengue fever certainly fit the predictions, the cause of that growth cannot be traced conclusively to global warming. Other factors could have been involved as well—for instance, disruption of the environment in ways that favor mosquito proliferation, declines in mosquito-control and other public health programs, and rises in drug and pesticide resistance. The case for a climatic contribution becomes stronger, however, when other projected consequences of global warming appear in concert with disease outbreaks.

Such is the case in highlands around the world. There, as anticipated, warmth is climbing up many mountains, along with plants and butterflies, and summit glaciers are melting. Since 1970 the elevation at which temperatures are always below freezing has ascended almost 500 feet in the tropics. Marching upward, too, are mosquitoes and mosquito-borne diseases.

In the 19th century, European colonists in Africa settled in the cooler mountains to escape the dangerous swamp air ("*mal aria*") that fostered disease in the lowlands. Today many of those havens are compromised. Insects and insect-borne infections are being reported at high elevations in South and Central America, Asia, and east and central Africa. Since 1980 *Ae. aegypti* mosquitoes, once limited by temperature thresholds to low altitudes, have been found above one mile in the highlands of northern India and at 1.3 miles in the Colombian Andes. Their presence magnifies the risk that dengue and yellow fever may follow. Dengue fever itself has struck at the mile mark in Taxco, Mexico. Patterns of insect migration change faster in the mountains than they do at sea level. Those alterations can thus serve as indicators of climate change and of diseases likely to expand their range.

Changes Are Already Under Way

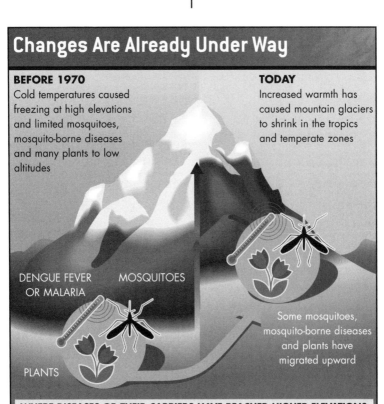

BEFORE 1970
Cold temperatures caused freezing at high elevations and limited mosquitoes, mosquito-borne diseases and many plants to low altitudes

TODAY
Increased warmth has caused mountain glaciers to shrink in the tropics and temperate zones

DENGUE FEVER OR MALARIA   MOSQUITOES

Some mosquitoes, mosquito-borne diseases and plants have migrated upward

PLANTS

WHERE DISEASES OR THEIR CARRIERS HAVE REACHED HIGHER ELEVATIONS

Malaria	Dengue fever	*Aedes aegypti* mosquitoes
Highlands of Ethiopia, Rwanda, Uganda and Zimbabwe Usamabara Mountains, Tanzania Highlands of Papua New Guinea and West Papua (Irian Jaya)	San Jose, Costa Rica Taxco, Mexico	(can spread dengue fever and yellow fever) Eastern Andes Mountains, Colombia Northern highlands of India

Computer models have predicted that global warming would produce several changes in the highlands: summit glaciers (like North Polar sea ice) would begin to melt, and plants, mosquitoes and mosquito-borne diseases would migrate upward into regions formerly too cold for them (*diagram*). All these predictions are coming true. This convergence strongly suggests that the upward expansion of mosquitoes and mosquito-borne diseases documented in the past 15 years has stemmed, at least in part, from rising temperatures.

Opportunists Like Sequential Extremes

The increased climate variability accompanying warming will probably be more important than the rising heat itself in fueling unwelcome outbreaks of certain vector-borne illnesses. For instance, warm winters followed by hot, dry summers (a pattern that could become all too familiar as the atmosphere heats up) favor the transmission of St. Louis encephalitis and other infections that cycle among birds, urban mosquitoes and humans.

This sequence seems to have abetted the surprise emergence of the West Nile virus in New York City last year. No one knows how this virus found its way into the U.S. But one reasonable explanation for its persistence and amplification here centers on the weather's effects on *Culex pipiens* mosquitoes, which accounted for the bulk of the transmission. These urban dwellers typically lay their eggs in damp basements, gutters, sewers and polluted pools of water.

The interaction between the weather, the mosquitoes and the virus probably went something like this: The mild winter of 1998–99 enabled many of the mosquitoes to survive into the spring, which arrived early. Drought in spring and summer concentrated nourishing organic matter in their breeding areas and simultaneously killed off mosquito predators, such as lacewings and ladybugs, that would otherwise have helped limit mosquito populations. Drought would also have led birds to congregate more, as they shared fewer and smaller watering holes, many of which were frequented, naturally, by mosquitoes.

Once mosquitoes acquired the virus, the heat wave that accompanied the drought would speed up viral maturation inside the insects. Consequently, as infected mosquitoes sought blood meals, they could spread the virus to birds at a rapid clip. As bird after bird became infected, so did more mosquitoes, which ultimately fanned out to infect human beings. Torrential rains toward the end of August provided new puddles for the breeding of *C. pipiens* and other mosquitoes, unleashing an added crop of potential virus carriers.

Like mosquitoes, other disease-conveying vectors tend to be "pests"—opportunists that reproduce quickly and thrive under disturbed conditions unfavorable to species with more specialized needs. In the 1990s climate variability contributed to the appearance in humans of a new rodent-borne ailment: the hantavirus pulmonary syndrome, a highly lethal infection of the lungs. This infection can jump from animals to humans when people inhale viral particles hiding in the secretions and excretions of rodents. The sequential weather extremes that set the stage for the first human eruption, in the U.S. Southwest in 1993, were long-lasting drought interrupted by intense rains.

First, a regional drought helped to reduce the pool of animals that prey on rodents—raptors (owls, eagles, prairie falcons, red-tailed hawks and kestrels), coyotes and snakes. Then, as drought yielded to unusually

El Niño's Message

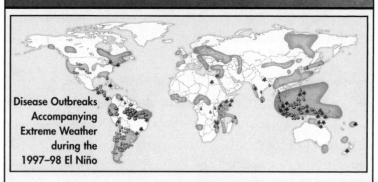

Disease Outbreaks
Accompanying
Extreme Weather
during the
1997–98 El Niño

Disease Outbreaks

Mosquito-borne: Dengue fever Rodent-borne: Hantavirus pulmonary syndrome
Encephalitis Waterborne: Cholera
Malaria Noninfectious: Respiratory illness resulting
Rift Valley fever from fire and smoke

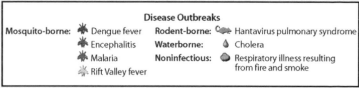

Extreme Weather
Abnormally wet areas
Abnormally dry areas

Scientists often gain insight into the workings of complicated systems by studying subsystems. In that spirit, investigators concerned about global warming's health effects are assessing outcomes of the El Niño Southern Oscillation (ENSO), a climate process that produces many of the same meteorological changes predicted for a warming world. The findings are not reassuring.

"El Niño" refers to an oceanic phenomenon that materializes every five years or so in the tropical Pacific. The ocean off Peru becomes unusually warm and stays that way for months before returning to normal or going to a cold extreme (La Niña). The name "Southern Oscillation" refers to atmospheric changes that happen in tandem with the Pacific's shifts to warmer or cooler conditions.

During an El Niño, evaporation from the heated eastern Pacific can lead to abnormally heavy rains in parts of South America and Africa and parts of Southeast Asia and Australia suffer droughts. Atmospheric pressure changes over the tropical Pacific also have ripple effects throughout the globe, generally yielding milder winters in some northern regions of the U.S. and western Canada. During a La Niña, weather patterns in the affected areas may go to opposite extremes.

The incidence of vector-borne and water-borne diseases climbs during El Niño and La Niña years, especially in areas hit by floods or droughts. Long-term studies in Colombia, Venezuela, India and Pakistan reveal, for instance, that malaria surges in the wake of El Niños. And my colleagues and I at Harvard University have shown that regions stricken by flooding or drought during the El Niño of 1997–98 (the strongest of the century) often had to contend as well with a convergence of diseases borne by mosquitoes, rodents and water (*map*). Additionally, in many dry areas, fires raged out of control, polluting the air for miles around.

ENSO is not merely a warning of troubles to come; it is likely to be an engine for those troubles. Several climate models predict that as the atmosphere and oceans heat up, El Niños themselves will become more common and

continued on following page

---

continued from previous page

severe—which means that the weather disasters they produce and the diseases they promote could become more prevalent as well.

Indeed, the ENSO pattern has already begun to change. Since 1976 the intensity, duration and pace of El Niños have increased. And during the 1990s, every year was marked by an El Niño or La Niña extreme. Those trends bode ill for human health in the 21st century.   —*P. R. E.*

---

heavy rains early in 1993, the rodents found a bounty of food, in the form of grasshoppers and piñon nuts. The resulting population explosion enabled a virus that had been either inactive or isolated in a small group to take hold in many rodents. When drought returned in summer, the animals sought food in human dwellings and brought the disease to people. By fall 1993, rodent numbers had fallen, and the outbreak abated.

Subsequent episodes of hantavirus pulmonary syndrome in the U.S. have been limited, in part because early-warning systems now indicate when rodent-control efforts have to be stepped up and because people have learned to be more careful about avoiding the animals' droppings. But the disease has appeared in Latin America, where some ominous evidence suggests that it may be passed from one person to another.

As the natural ending of the first hantavirus episode demonstrates, ecosystems can usually survive occasional extremes. They are even strengthened by seasonal changes in weather conditions, because the species that live in changeable climates have to evolve an ability to cope with a broad range of conditions. But long-lasting extremes and very wide fluctuations in weather can overwhelm ecosystem resilience. (Persistent ocean heating, for instance, is menacing coral reef systems, and drought-driven forest fires are threatening forest habitats.) And ecosystem upheaval is one of the most profound ways in which climate change can affect human health. Pest control is one of nature's underappreciated services to people; well-functioning ecosystems that include diverse species help to keep nuisance organisms in check. If increased warming and weather extremes result in more ecosystem disturbance, that disruption may foster the growth of opportunist populations and enhance the spread of disease.

Unhealthy Water

Beyond exacerbating the vector-borne illnesses mentioned above, global warming will probably elevate the incidence of waterborne diseases, including cholera (a cause of severe diarrhea). Warming itself can contribute to the change, as can a heightened frequency and extent of droughts and floods. It may seem strange that droughts would favor waterborne disease, but they can wipe out supplies of safe drinking water and concentrate contaminants that might otherwise remain dilute. Further, the lack of clean water during a drought interferes with good hygiene and safe rehydration of those who have lost large amounts of water because of diarrhea or fever.

Floods favor waterborne ills in different ways. They wash sewage and other sources of pathogens (such as *Cryptosporidium*) into supplies of drinking water. They also flush fertilizer into water supplies. Fertilizer and sewage can each combine with warmed water to trigger expansive blooms of harmful algae. Some of these blooms are directly toxic to humans who inhale their vapors; others contaminate fish and shell-fish, which, when eaten, sicken the consumers. Recent discoveries have revealed that algal blooms can threaten human health in yet another way. As they grow bigger, they support the proliferation of various pathogens, among them *Vibrio cholerae*, the causative agent of cholera.

Drenching rains brought by a warmed Indian Ocean to the Horn of Africa in 1997 and 1998 offer an example of how people will be affected as global warming spawns added flooding. The downpours set off epidemics of cholera as well as two mosquito-borne infections: malaria and Rift Valley fever (a flu-like disease that can be lethal to livestock and people alike).

To the west, Hurricane Mitch stalled over Central America in October 1998 for three days. Fueled by a heated Caribbean, the storm unleashed torrents that killed at least 11,000 people. But that was only the beginning of its havoc. In the aftermath, Honduras reported thousands of cases of cholera,

Weather and the West Nile Virus

This diagram offers a possible explanation for how a warming trend and sequential weather extremes helped the West Nile virus to establish itself in the New York City area in 1999. Whether the virus entered the U.S. via mosquitoes, birds or people is unknown. But once it arrived, interactions between mosquitoes and birds amplified its proliferation.

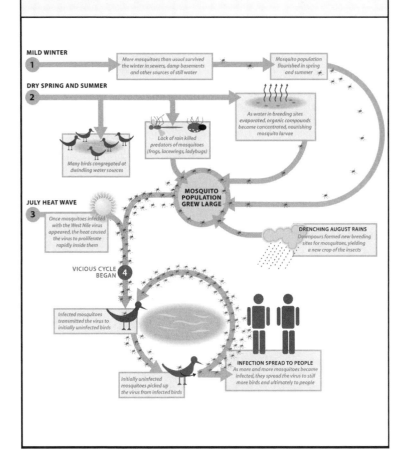

MILD WINTER

1 → More mosquitoes than usual survived the winter in sewers, damp basements and other sources of still water → Mosquito population flourished in spring and summer

DRY SPRING AND SUMMER

2

As water in breeding sites evaporated, organic compounds became concentrated, nourishing mosquito larvae

Lack of rain killed predators of mosquitoes (frogs, lacewings, ladybugs)

Many birds congregated at dwindling water sources

MOSQUITO POPULATION GREW LARGE

JULY HEAT WAVE

3 Once mosquitoes infected with the West Nile virus appeared, the heat caused the virus to proliferate rapidly inside them

DRENCHING AUGUST RAINS Downpours formed new breeding sites for mosquitoes, yielding a new crop of the insects

VICIOUS CYCLE BEGAN 4

Infected mosquitoes transmitted the virus to initially uninfected birds

INFECTION SPREAD TO PEOPLE As more and more mosquitoes became infected, they spread the virus to still more birds and ultimately to people

Initially uninfected mosquitoes picked up the virus from infected birds malaria and dengue fever. Beginning in February of this year, unprecedented rains and a series of cyclones inundated large parts of southern Africa. Floods in Mozambique and Madagascar killed hundreds, displaced thousands and spread both cholera and malaria. Such events can also greatly retard economic development, and its accompanying public health benefits, in affected areas for years.

Solutions

The health toll taken by global warming will depend to a large extent on the steps taken to prepare for the dangers. The ideal defensive strategy would have multiple components.

One would include improved surveillance systems that would promptly spot the emergence or resurgence of infectious diseases or the vectors that carry them. Discovery could quickly trigger measures to control vector proliferation without harming the environment, to advise the public about self-protection, to provide vaccines (when available) for at-risk populations and to deliver prompt treatments.

This past spring, efforts to limit the West Nile virus in the northeastern U.S. followed this model. On seeing that the virus had survived the winter, public health officials warned people to clear their yards of receptacles that can hold stagnant water favorable to mosquito breeding. They also introduced fish that eat mosquito larvae into catch basins and put insecticide pellets into sewers.

Sadly, however, comprehensive surveillance plans are not yet realistic in much of the world. And even when vaccines or effective treatments exist, many regions have no means of obtaining and distributing them. Providing these preventive measures and treatments should be a global priority.

A second component would focus on predicting when climatological and other environmental conditions could become conducive to disease outbreaks, so that the risks could be minimized. If climate models indicate that floods are likely in a given region, officials might stock shelters with extra supplies. Or if satellite images and sampling of coastal waters indicate that algal blooms related to cholera outbreaks are beginning, officials could warn people to filter contaminated water and could advise medical facilities to arrange for additional staff, beds and treatment supplies.

Research reported in 1999 illustrates the benefits of satellite monitoring. It showed that satellite images detecting heated water in two specific ocean regions and lush vegetation in the Horn of Africa can predict outbreaks of Rift Valley fever in the Horn five months in advance. If such assessments led to vaccination campaigns in animals, they could potentially forestall epidemics in both livestock and people.

A third component of the strategy would attack global warming itself. Human activities that contribute to the heating or that exacerbate its effects must be limited. Little doubt remains that burning fossil fuels for energy is playing a significant role in global warming, by spewing carbon dioxide and other heat-absorbing, or "greenhouse," gases into the air. Cleaner energy sources must be put to use quickly and broadly, both in the energy-guzzling industrial world and in developing nations, which cannot be expected to cut back on their energy use. (Providing sanitation, housing, food, refrigeration and indoor fires for cooking takes energy, as do the pumping and purification of water and the desalination of seawater for irrigation.) In parallel, forests and wetlands need to be restored, to absorb carbon dioxide and floodwaters and to filter contaminants before they reach water supplies.

The world's leaders, if they are wise, will make it their business to find a way to pay for these solutions. Climate, ecological systems and society can all recoup after stress, but only if they are not exposed to prolonged challenge or to one disruption after another. The Intergovernmental Panel on Climate Change, established by the United Nations, calculates that halting the ongoing rise in atmospheric concentrations of greenhouse gases will require a whopping 60 to 70 percent reduction in emissions.

I worry that effective corrective measures will not be instituted soon enough. Climate does not necessarily change gradually. The multiple factors that are now destabilizing the global climate system could cause it to jump abruptly out of its current state. At any time, the world could suddenly become much hotter or even much colder. Such a sudden, catastrophic change is the ultimate health risk—one that must be avoided at all costs.

The Author

Paul R. Epstein, an M.D. trained in tropical public health, is associate director of the Center for Health and the Global Environment at Harvard Medical School. He has served in medical, teaching and research capacities in Africa, Asia and Latin America and has worked with the Intergovernmental Panel on Climate Change, the National Oceanic and Atmospheric Administration, and the National Aeronautics and Space Administration to assess the health effects of climate change and to develop health applications for climate forecasting and remote-sensing technologies.

Thomas R. Karl and Kevin E. Trenberth, two major players in climate research, here summarize many of the basic issues of the field, from scientific and political matters to sophisticated computer forecasting models and basic human choices. It may be helpful, before reading this, to think about three core types of climate research, each of which is necessary in the effort to paint a complete picture of the earth's climatological past while also predicting its future: computer modeling, historical reconstruction of climates, and monitoring.

The rapidly increasing power of computers will surely help climatologists, who have developed such unfathomably complex computer models that their programs can take days to arrive at a prediction for a mere month into the climatological future. To construct climate histories and sort out some of the long-term connections between things like greenhouse gases and temperature, scientists study bubbles of gas trapped in ancient ice, pulled up in deep cores from places like Greenland. They peer into the growth rings of corals and trees, whose growth reflects local climates, and they study the hard remains of tiny ocean organisms pulled from the densely layered seafloor. To monitor the details of modern climate, scientists use satellites and ocean buoys and other equipment. Karl and Trenberth worry that the U.S. government has not committed enough time, money, or resources to the kind of consistent, continual climate modeling that will help reveal what our climatological future holds. —KH

"The Human Impact on Climate"
by Thomas R. Karl and Kevin E. Trenberth
Scientific American, December 1999

"The balance of evidence suggests a discernible human influence on global climate." With these carefully chosen words, the Intergovernmental Panel on Climate Change (jointly supported by the World Meteorological

Organization and the United Nations Environmental Program) recognized in 1995 that human beings are far from inconsequential when it comes to the health of the planet. What the panel did not spell out—and what scientists and politicians dispute fiercely—is exactly when, where and how much that influence has and will be felt.

So far the climate changes thought to relate to human endeavors have been relatively modest. But various projections suggest that the degree of change will become dramatic by the middle of the 21st century, exceeding anything seen in nature during the past 10,000 years. Although some regions may benefit for a time, overall the alterations are expected to be disruptive or even severe. If researchers could clarify the extent to which specific activities influence climate, they would be in a much better position to suggest strategies for ameliorating the worst disturbances. Is such quantification possible? We think it is and that it can be achieved by the year 2050—but only if that goal remains an international priority.

Despite uncertainties about details of climate change, our activities clearly affect the atmosphere in several troubling ways. Burning of fossil fuels in power plants and automobiles ejects particles and gases that alter the composition of the atmosphere. Visible pollution from sulfur-rich fuels includes micron-size particles called aerosols, which often cast a milky haze in the sky. These aerosols temporarily cool the atmosphere because they reflect some of the sun's rays back to space, but they stay in the air for only a few days before rain sweeps them to the planet's surface. Certain invisible gases deliver a more lasting impact. Carbon dioxide remains in the atmosphere for a century or more. Worse yet, such greenhouse gases trap some of the solar radiation that the planet would otherwise radiate back to space, creating a "blanket" that insulates and warms the lower atmosphere.

Indisputably, fossil-fuel emissions alone have increased carbon dioxide concentrations in the atmosphere by about 30 percent since the start of the Industrial Revolution in the late 1700s. Oceans and plants help to offset this flux by scrubbing some of the gas out of the air over time, yet carbon dioxide concentrations continue to grow. The inevitable result of pumping the sky full of greenhouse gases is global warming. Indeed, most scientists agree that the earth's mean temperature has risen at least 0.6 degree Celsius (more than one degree Fahrenheit) over the past 120 years, much of it caused by the burning of fossil fuels.

The global warming that results from the greenhouse effect dries the planet by evaporating moisture from oceans, soils and plants. Additional moisture in the atmosphere provides a swollen reservoir of water that is tapped by all precipitating weather systems, be they tropical storms, thundershowers, snowstorms or frontal systems. This enhanced water cycle brings on more severe droughts in dry areas and leads to strikingly heavy rain or snowfall in wet regions, which heightens the risk of flooding. Such weather patterns have burdened many parts of the world in recent decades.

Human activities aside from burning fossil fuels can also wreak havoc on the climate system. For instance, the conversion of forests to farmland eliminates trees that would otherwise absorb carbon from the atmosphere and reduce the greenhouse effect. Fewer trees also mean greater rainfall runoff, thereby increasing the risk of floods.

It is one thing to have a sense of the factors that can bring about climate change. It is another to know how the human activity in any given place will affect the local and global climate. To achieve that aim, those of us who are concerned about the human influence on climate will have to be able to construct more accurate climate models than have ever been designed before. We will therefore require the technological muscle of supercomputers a million times faster than those in use today. We will also have to continue to disentangle the myriad interactions among the oceans, atmosphere and biosphere to know exactly what variables to feed into the computer models.

Most important, we must be able to demonstrate that our models accurately simulate past and present climate change before we can rely on models to predict the future. To do that, we need long-term records. Climate simulation and prediction will come of age only with an ongoing record of changes as they happen.

Computers and Climate Interactions

For scientists who model climate patterns, everything from the waxing and waning of ice ages to the desertification of central Africa plays out inside the models run on supercomputers. Interactions among the components of the climate system—the atmosphere, oceans, land, sea ice, freshwater and biosphere—behave according to physical laws represented by dozens of mathematical equations. Modelers instruct the computers to solve these equations for each box in a three-dimensional grid that covers the globe. Because nature is not constrained by boxes, the chore is not only to incorporate the correct mathematics within each box but also to describe appropriately the transfer of energy and mass into and out of the boxes.

The computers at the world's preeminent climate-modeling facilities can perform between 10 and 50 billion operations per second, but with so many evolving variables, the simulation of a single century can take months. The time it takes to run a simulation, then, limits the resolution (or number of boxes) that can be included within climate models. For typical models designed to mimic the detailed evolution of weather systems, boxes in the three-dimensional grid measure about 250 kilometers (156 miles) square in the horizontal direction and one kilometer in the vertical. Tracking patterns within smaller areas thus proves especially difficult.

Even the most sophisticated of our current global models cannot directly simulate conditions such as cloud cover and the formation of rain. Powerful thunderstorm clouds that can unleash sudden downpours often operate on scales of less than 10 kilometers, and raindrops condense at submillimeter scales. Because each of these events happens in a region smaller than the volume of the smallest grid unit, their characteristics must be inferred by elaborate statistical techniques.

Such small-scale weather phenomena develop randomly. The frequency of these random events can differ extensively from place to place, but most agents that alter climate, such as rising levels of greenhouse gases, affect all areas of the planet much more uniformly. The variability of weather will increasingly mask large-scale climate activity as smaller regions are considered. Lifting that mask thus drains computer time, because it requires running several simulations, each with slightly different starting conditions. The climate features that occur in every simulation constitute the climate "signal," whereas those that are not reproducible are considered weather-related climate "noise."

Conservative estimates indicate that computer-processing speed will have increased by well over a million times by 2050. With that computational power, climate modelers could perform many simulations with different starting conditions and better distinguish climate signals from climate noise. We could also routinely run longer simulations of hundreds of years with less than one-kilometer horizontal resolution and an average of 100-meter vertical resolution over the oceans and atmosphere.

Faster computers help to predict climate change only if the mathematical equations fed into them perfectly describe what happens in nature. For example, if a model atmosphere is simulated to be too cold by four degrees C (not uncommon a decade ago), the simulation will indicate that the atmosphere can hold about 20 percent less water than its actual capacity— a significant error that renders meaningless any subsequent estimates of evaporation and precipitation. Another problem is that we do not yet know how to replicate adequately all the processes that influence climate, such as hiccups in the carbon cycle and modifications in land use. What is more, these changes can initiate feedback cycles that, if ignored, can lead the model astray. Raising temperature, for example, sometimes enhances another variable, such as moisture content of the atmosphere, which in turn amplifies the original perturbation. (In this case, more moisture in the air causes increased warming because water vapor is a powerful greenhouse gas.)

Researchers are only beginning to realize how much some of these positive feedbacks influence the planet's life-giving carbon cycle. The 1991 eruption of Mount Pinatubo in the Philippines, for instance, belched out enough ash and sulfur dioxide to cause a temporary global cooling as those compounds interacted with water droplets in the air to block some of the sun's incoming radiation. This depleted energy can inhibit carbon dioxide uptake in plants.

Using land in a different way can perturb continental and regional climate systems in ways that are difficult to translate into equations. Clearing forests for farming and ranching brightens the land surface. Croplands are lighter-colored than dark forest and thus reflect more solar radiation, which tends to cool the atmosphere, especially in autumn and summer.

Dearth of Data

Climate simulations can never move out of the realm of good guesses without accurate observations to validate them and to show that the models do indeed reflect reality. In other words, to reduce our uncertainty about the sensitivity of the climate system to human activity, we need to know how the climate has changed in the past. We must be capable of adequately simulating conditions before the Industrial Revolution and especially since that time, when humans have altered irrevocably the composition of the atmosphere.

To understand climate from times prior to the development of weather-tracking satellites and other instruments, we rely on indicators such as air and chemicals trapped in ice cores, the width of tree rings, coral growth, and sediment deposits on the bottoms of oceans and lakes. These snapshots provide us with information that aids in piecing together past conditions. To truly understand the present climate, however, we require more than snapshots of physical, chemical and biological quantities; we also need the equivalent of long-running videotape records of the currently evolving climate. Ongoing measurements of sea ice, snow cover, soil moisture, vegetative cover, and ocean temperature and salinity are just some of the variables involved.

But the present outlook is grim: no U.S. or international institution has the mandate or resources to monitor long-term climate. Scientists currently compile their interpretations of climate change from large networks of satellites and surface sensors such as buoys, ships, observatories, weather stations and airplanes that are being operated for other purposes, such as short-term weather forecasting. As a result, depictions of past climate variability are often equivocal or missing.

The National Oceanic and Atmospheric Administration operates many of these networks, but it does not have the resources to commit to a long-term climate-monitoring program. Even the National Aeronautics and Space Administration's upcoming Earth Observing System, which entails launching several sophisticated satellites to monitor various aspects of global systems, does not include the continuity of a long-term climate observation program in its mission statement.

Whatever the state of climate monitoring may be, another challenge in the next decade will be to ensure that the quantities we do measure actually represent real multidecadal changes in the environment. In other words, what happens if we use a new camera or point it in a different direction? For instance, a satellite typically lasts only four years or so before it is replaced with another in a different orbit. The replacement usually has new instruments and observes the earth at a different time of day. Over a period of years, then, we end up measuring not only climate variability but also the changes introduced by observing the climate in a different way. Unless precautions are taken to quantify the modifications in observing technology and sampling methods before the older technology is replaced, climate records could be rendered useless because it will be impossible to compare the new set of data with its older counterpart.

Future scientists must be able to evaluate their climate simulations with unequivocal data that are properly archived. Unfortunately, the data we have archived from satellites and critical surface sensors are in jeopardy of being lost forever. Longterm surface observations in the U.S. are still being recorded on outdated punched paper tapes or are stored on decaying paper or on old computer hardware. About half the data from our new Doppler radars are lost because the recording system relies on people to deal with the details of data preservation during severe weather events, when warnings and other critical functions are a more immediate concern.

Can We Realize the Vision?

Over the next 50 years we can broadly understand, if we choose to, how human beings are affecting the global, regional and even small-scale aspects of climate. But waiting until then to take action would be fool-hardy. Long lifetimes of carbon dioxide and other greenhouse gases in the atmosphere, coupled with the climate's typically slow response to evolving conditions, mean that even if we cut back on harmful human activities today, the planet very likely will still undergo substantial change.

Glaciers melting in the Andes highlands and elsewhere are already confirming the reality of a warming planet. Rising sea level—and drowning coastlines—testify to the projected global warming of perhaps two degrees C or more by the end of the next century. Climate change will in all likelihood capture the most attention when its effects exacerbate other pressures on society. The spread of settlements into coastal regions and low-lying areas vulnerable to flooding is just one of the initial difficulties that we will most likely face. But as long as society can fall back on the uncertainty of human impact on climate, legislative mandates for changing standards of fossil-fuel emissions or forest clear-cutting will be hard fought.

The need to foretell how much we influence our world argues for doing everything we can to develop comprehensive observing and data-archiving systems now. The resulting information could feed models that help make skillful predictions of climate several years in advance. With the right planning we could be in a position to predict, for example, exactly how dams and reservoirs might be better designed to accommodate anticipated floods and to what extent green house gas emissions from new power plants will warm the planet.

Climate change is happening now, and more change is certain. We can act to slow it down, and we can sensibly plan for it, but at present we are doing neither. To anticipate the true shape of future climate, scientists must overcome the obstacles we have outlined above. The need for greater computer power and for a more sophisticated understanding of the nuances of climate interactions should be relatively easy to overcome. The real stumbling block is the long-term commitment to global climate monitoring. How can we get governments to commit resources for decades of surveys, particularly when so many governments change hands with such frequency?

If we really want the power to predict the effects of human activity by 2050—and to begin addressing the disruption of our environment—we must pursue another path. We have a tool to clear such a path: the United Nations Framework Convention on Climate Change, signed by President George Bush in 1992. The convention binds together 179 governments with a commitment to remedy damaging human influence on global climate. The alliance took a step toward stabilizing greenhouse gas emissions by producing the Kyoto Protocol in 1997, but long-term global climate-monitoring systems remain unrealized.

The Authors

Thomas R. Karl has directed the National Climatic Data Center (NCDC) in Asheville, N.C., since March 1998. The center is part of the National Oceanic and Atmospheric Administration and serves as the world's largest active archive of climate data. Karl, who has worked at the center since 1980, has focused much of his research on climate trends and extreme weather. He also writes reports for the Intergovernmental Panel on Climate Change (IPCC), the official science source for international climate change negotiations.

Kevin E. Trenberth directs the Climate Analysis section at the National Center for Atmospheric Research (NCAR) in Boulder, Colo., where he studies El Niño and climate variability. After several years in the New Zealand Meteorological Service, he became a professor of atmospheric sciences at the University of Illinois in 1977 and moved to NCAR in 1984. Trenberth also co-writes IPCC reports with Karl.

The Ongoing Debate over Climate Change

Michael Mann's research, widely accepted in the scientific community, is less universally accepted in political arenas, where the science of climate change is often labeled "uncertain." That accusation is certainly true on some level: Researchers still cannot accurately predict whether Washington, D.C., will see more rain or more drought in the future, or if Alaska's glaciers will be gone by 2050 or 2150. However, most scientists accept that climate change is already well under way and will continue. They feel the scientific debate is now beyond question and at an end. The remaining issues to discuss venture beyond science into economic and social policy. Given that climates are changing, what should we do about it? Would the economy be imperiled if we took drastic action to curb industrial pollution and the burning of fossil fuels? Would it be catastrophically costly if we failed to take action and unchecked climate change adversely affected weather, agriculture, and human population centers?

This article provides a portrait of a climate researcher who has weathered wave after wave

*of blistering political attacks for his politically
unpopular conclusions. Soon after it was written,
Congressman Joe Barton, a Republican from
Texas, launched an investigation of Mann's
research. The investigation was quickly attacked
by other politicians—primarily by another
prominent Republican member of Congress from
New York—as politically motivated and designed
to silence criticism of prevailing government
policy on issues of climate, industry, energy,
and the environment. Even the National
Academy of Sciences weighed in, offering to
convene an independent panel to assess the
scientific consensus on climate change.
Representative Barton declined an invitation to
participate in the panel's work. —KH*

"Behind the Hockey Stick"
by David Appell
Scientific American, March 2005

Michael Mann knows his students and his subject. The
topic of the graduate seminar: El Niño and radiative
forcing. The beer he will be serving: Corona, "because
I'm going to be talking about tropical climate." Not
surprisingly, attendance is high.

Mann is most famously known for the "hockey
stick," a plot of the past millennium's temperature
that shows the drastic influence of humans in the 20th
century. Specifically, temperature remains essentially

flat until about 1900, then shoots up, like the upturned blade of a hockey stick. The work was also the first to add error bars to the historical temperatures and allow for regional reconstructions of temperature.

That stick has become a focal point in the controversy surrounding climate change and what to do about it. Proponents see it as a clear indicator that humans are warming the globe; skeptics argue that the climate is undergoing a natural fluctuation not unlike those in eras past. But Mann has not been deterred by the attacks. "If we allowed that sort of thing to stop us from progressing in science, that would be a very frightening world," says the 39-year-old climatologist in his University of Virginia office overlooking the hills of Monticello, the home of Thomas Jefferson.

To construct the hockey-stick plot, Mann, Raymond S. Bradley of the University of Massachusetts Amherst and Malcolm K. Hughes of the University of Arizona analyzed paleoclimatic data sets such as those from tree rings, ice cores and coral, joining historical data with thermometer readings from the recent past. In 1998 they obtained a "reconstruction" of Northern Hemisphere temperatures going back 600 years; by the next year they had extended their analysis to the past 1,000 years. In 2003 Mann and Philip D. Jones of the University of East Anglia in England used a different method to extend results back 2,000 years.

In each case, the outcome was clear: global mean temperature began to rise dramatically in the early 20th century. That rise coincided with the unprecedented release of carbon dioxide and other heat-trapping gases into the earth's atmosphere, leading to the conclusion that industrial activity was boosting the world's mean temperature. Other researchers subsequently confirmed the plot.

The work of Mann and his colleagues achieved special prominence in 2001. That is when the Intergovernmental Panel on Climate Change (IPCC), an international body of climate experts, placed the hockey-stick chart in the Summary for Policymakers section of the panel's Third Assessment Report. (Mann also co-authored one of the chapters in the report.) It thereby elevated the hockey stick to iconic status—as well as making it a bull's-eye. A community skeptical of human-induced warming argued that Mann's data points were too sparse to constitute a true picture, or that his raw data were numerically suspicious, or that they could not reproduce his results with the data he had used. Take down Mann, it seemed, and the rest of the IPCC's conclusions about anthropogenic climate change would follow.

That led to "unjustified attack after unjustified attack," complains climatologist Gavin A. Schmidt of the NASA Goddard Institute for Space Studies. Although questions in the field abound about how, for example, tree-ring data are compiled, many of those attacking Mann's work, Schmidt claims, have had a priori opinions that the work must be wrong. "Most scientists would have left the field long ago, but Mike is fighting back with a tenacity I find admirable,"

Schmidt says. One of Mann's more public punch backs took place in July 2003, when he defended his views before a congressional committee led by Senator James M. Inhofe of Oklahoma, who has called global warming a "hoax." "I left that meeting having demonstrated what the mainstream views on climate science are," Mann asserts.

More recently, Mann battled back in a 2004 corrigendum in the journal *Nature*, in which he clarified the presentation of his data. He has also shown how errors on the part of his attackers led to their specific results. For instance, skeptics often cite the Little Ice Age and Medieval Warming Period as pieces of evidence not reflected in the hockey stick, yet these extremes are examples of regional, not global, phenomena. "From an intellectual point of view, these contrarians are pathetic, because there's no scientific validity to their arguments whatsoever," Mann says. "But they're very skilled at deducing what sorts of disingenuous arguments and untruths are likely to be believable to the public that doesn't know better."

Mann thinks that the attacks will continue, because many skeptics, such as the Greening Earth Society and the Tech Central Station Web site, obtain funds from petroleum interests. "As long as they think it works and they've got unlimited money to perpetuate their disinformation campaign," Mann believes, "I imagine it will go on, just as it went on for years and years with tobacco until it was no longer tenable—in fact, it became perjurable to get up in a public forum

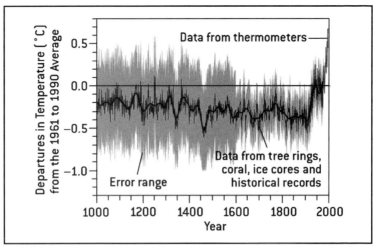

"Hockey Stick" graph shows a 20th-century upturn in temperature in the Northern Hemisphere. The error range is greater in the past because the data are sparser.

and claim that there was no science" behind the health hazards of smoking.

As part of his hockey-stick defense, Mann co-founded with Schmidt a Weblog called RealClimate (www.realclimate.org). Started in December 2004, the site has nine active scientists, who have attracted the attention of the blog cognoscenti for their writings, including critiques of Michael Crichton's *State of Fear*, a novel that uses charts and references to argue against anthropogenic warming. The blog is not a bypass of the ordinary channels of scientific communication, Mann explains, but "a resource where the public can go to see what actual scientists working in the field have to say about the latest issues."

The most challenging aspect today, Mann thinks, is predicting regional disruptions, because people are unlikely to take climate change seriously until they see how it operates in their backyard. In that regard, he has turned his attention to El Niño, a warming of eastern tropical Pacific waters that affects global weather. In discussing the issue with his students over their Coronas, Mann notes that comparisons with the paleoclimatic record seem to confirm a mechanism proposed by other researchers. Specifically, radiative forcings—volcanic eruptions and solar changes, for instance—do in fact alter El Niño, turning it into more of a La Niña state, with colder sea-surface temperatures. Understanding how El Niño has changed with past radiative forcings is a first step to understanding how it will change in an increasingly greenhouse-gassed world.

Mann remains somewhat mum on whether the U.S. should join the Kyoto Treaty, an international agreement to limit fossil-fuel emissions: "It's hard enough predicting the climate. I don't pretend to be able to predict the behavior of politicians." He sees the Kyoto accord as an initial step that is unlikely to curtail emissions all that much, but it will at least set in motion a process that can be built on with other treaties.

Such efforts are essential, because the blade of Mann's hockey stick will get longer. He notes that "we're already committed to 50 to 100 years of warming and several centuries of sea-level rise, simply from the amount of greenhouse gases we've already put in the atmosphere." The solution to global warming, he observes, "is going to be finding an appropriate set of constraints on fossil-fuel emissions that allow us to slow the rate of change down to a level we can adapt to."

David Appell is based in Newmarket, N.H.

A strange thing happened in 1991. The level of the greenhouse gas carbon dioxide dropped in the atmosphere, after more than four decades of increase. Temperatures dipped, too. Global warming skeptics crowed that this was proof the world's climate was not, in fact, changing because of human activity. Scientists were not so sure.

It was relatively easy to attribute the temperature dip to the dusty eruption of Mount Pinatubo in the Philippines—the dust blocked out some sun for a while, cooling things down. It was harder, however, to explain the dip in carbon dioxide levels. Where did some of that gas go? Into the ocean? Into land? This article describes how scientists tried to solve the mystery of the missing carbon dioxide. —KH

"No Global Warming?"
by Kristin Leutwyler
Scientific American, **February 1994**

Since 1958, when researchers first began to measure the rate at which carbon dioxide accumulates in the atmosphere, they have seen a consistent increase, perturbed only by minor seasonal fluctuations. Then, about four years ago, the trend began to waver. First a decline set in, followed by a plateau. After that, the decline resumed—sharply. The event has left scientists, including those at the observatory of Mauna Loa in Hawaii, established by the late Harry Wexler to make the measurements, wondering what has happened.

Adding to the confusion, says Charles D. Keeling of the University of California at San Diego, who has operated a gas analyzer at the observatory since its founding, is the fact that accumulation started to slump while the atmosphere was in the throes of an El Niño, a periodic shift in the circulation of trade winds over the Pacific that affects global weather and ocean currents. During an El Niño, such as those of 1982–83 and 1986–87, atmospheric carbon dioxide levels tend to rise faster than they do at other times. Keeling suspects that plants and soils release more carbon dioxide during an El Niño because when an Asian monsoon collapses, it causes drought conditions. Whatever has been reducing contributions of carbon dioxide to the atmosphere had such an impact that it entirely overrode the effects of an El Niño.

Any number of events might have had such climatic clout. Scientists can eliminate only one explanation immediately: the amount of carbon dioxide released from burning fossil fuels has not declined. The next most obvious candidate is the June 1991 eruption of Mount Pinatubo in the Philippines. "The link to the eruption is pretty speculative, but it's an attractive thing to think about because of the coincidence in time," says Ralph F. Keeling, Charles Keeling's son and colleague at U.C.S.D. Of course, discovering whether the mystery source existed at land or at sea would narrow the search further. Unfortunately, different tests have yielded conflicting clues.

The ratio of carbon 13 to carbon 12 in the atmosphere is one such measure. Photosynthesis on land prefers the lighter isotope, whereas gas exchange at sea discriminates only slightly between the two. "We saw the ratio go up, which would imply an increased carbon dioxide uptake by the terrestrial biosphere," says Pieter P. Tans of the National Oceanic and Atmospheric Administration. "But there could be considerable error in that. It is very dependent on how good our calibration is." Indeed, researchers measuring the carbon isotope ratio have reported different results at various meetings over the past year. Charles Keeling's data initially indicated a large sink at sea. After corrections were made to his calibration, the results instead pointed to a sink predominantly on land.

Oxygen emissions, on the other hand, support yet another idea. "It's fairly clear that the land did not behave in a typical way for an El Niño, but the oxygen data suggest that maybe the oceans also behaved strangely," Ralph Keeling says. Just as different flavors of carbon isotopes are preferred by surf-and-turf reactions, so, too, varying proportions of oxygen and carbon are engaged through the formation and consumption of organic matter. In addition, carbon is quite reactive at sea, whereas oxygen is chemically neutral.

After considerable number crunching, these facts taken together imply that if the sink were primarily on land, as the carbon isotope readings suggest, the change in the growth rate of atmospheric oxygen should be nearly equivalent to the recent change for carbon dioxide. In fact, Ralph Keeling has observed oxygen emissions that rose about twice as sharply as the rate by which carbon dioxide emissions fell after the Pinatubo event. This finding indicates that significant changes took place in the oceans.

No matter where this carbon sink existed, scientists face the additional challenge of figuring out how it happened. There are several models based on the fallout from Pinatubo that could conceivably illustrate why carbon dioxide emissions plummeted. Global cooling, measured in the low troposphere via satellite, provides one compelling pathway. Such cooling could affect the balance between photosynthesis and respiration on land and could lead to an increased uptake of carbon dioxide in the oceans. "It could cause a big, short-term jolt to the carbon balance. In 1994, if the temperature comes back to normal, we should get normal carbon dioxide growth again," Tans notes.

So, is global warming on the way out? Tans does not think so. The decline in atmospheric carbon dioxide accumulation, he believes, is temporary. Ralph Keeling agrees. "That the carbon dioxide growth will stay low is doubtful," he says. "But this is relevant at least in the sense that it shows we don't really know what's happening with respect to the most important man-made greenhouse gas."

As David Appell reports in the following article, the stakes are increasingly high in debates about human-caused global warming. People are beginning to take note of climate changes, from melting glaciers to the increasing frequency and intensity of heat waves and droughts. Industries that depend on fossil fuels—oil and gas companies, in particular—stand to lose if both politicians and ordinary citizens begin to heed the calls to pump fewer greenhouse gases into the air.

For this reason, some industry groups trumpeted the research project described in this article. To understand the research of Willie Soon and Sallie Baliunas discussed in this article, it is important to understand the scientific technique of meta-analysis, that is, taking other researchers' studies and analyzing them in aggregate. In this case, the researchers classified scientific studies as either supporting or opposing the idea that two historic climate events were global, not regional. It will become clear as you read why the distinction between global and local climate events is important, but many researchers—even supporters of the paper's conclusion that the current period of global warming is neither unique nor extreme—take issue with Soon and Baliunas's methodology. —KH

"Hot Words"
by David Appell
Scientific American, August 2003

In a contretemps indicative of the political struggle over global climate change, a recent study suggested that humans may not be warming the earth. Greenhouse skeptics, pro-industry groups and political conservatives have seized on the results, proclaiming that the science of climate change is inconclusive and that agreements such as the Kyoto Protocol, which set limits on the output of industrial heat-trapping gases, are unnecessary. But mainstream climatologists, as represented by the Intergovernmental Panel on Climate Change (IPCC), are perturbed that the report has received so much attention; they say the study's conclusions are scientifically dubious and colored by politics.

Sallie Baliunas and Willie Soon of the Harvard-Smithsonian Center for Astrophysics reviewed more than 200 studies that examined climate "proxy" records—data from such phenomena as the growth of tree rings or coral, which are sensitive to climatic conditions. They concluded in the January [2003] *Climate Research* that "across the world, many records reveal that the 20th century is probably not the warmest nor a uniquely extreme climate period of the last millennium." They said that two extreme climate periods—the Medieval Warming Period between 800 and 1300 and the Little Ice Age of 1300 to 1900—occurred worldwide, at a time before industrial emissions of greenhouse gases became abundant. (A longer version subsequently appeared in the May [2003] *Energy and Environment.*)

In contrast, the consensus view among paleoclimatologists is that the Medieval Warming Period was regional, that the worldwide nature of the Little Ice Age is open to question and that the late 20th century saw the most extreme global average temperatures.

Scientists skeptical of human-induced warming applaud the analysis by Soon and Baliunas. "It has been painstaking and meticulous," says William Kininmonth, a meteorological consultant in Kew, Australia, and former head of the Australian National Climate Center. But he says that "from a purely statistical viewpoint, the work can be criticized."

And that criticism, from many scientists who feel that Soon and Baliunas produced deeply flawed work, has been unusually strident. "The fact that it has received any attention at all is a result, again in my view, of its utility to those groups who want the global warming issue to just go away," comments Tim Barnett, a marine physicist at the Scripps Institution of Oceanography, whose work Soon and Baliunas refer to. Similar sentiments came from Malcolm Hughes of the Laboratory of Tree-Ring Research at the University of Arizona, whose work is also discussed: "The Soon et al. paper is so fundamentally misconceived and contains so many egregious errors that it would take weeks to list and explain them all."

Rather than seeing global anomalies, many paleo-climatologists subscribe to the conclusions of Phil Jones of the University of East Anglia, Michael Mann of the University of Virginia and their colleagues, who began in 1998 to quantitatively splice together the proxy records. They have concluded that the global average temperature over the past 1,000 years has been relatively stable until the 20th century. "Nothing in the paper undermines in any way the conclusion of earlier studies that the average temperature of the late twentieth century in the Northern Hemisphere was anomalous against the background of the past millennium," wrote Mann and Princeton University's Michael Oppenheimer in a privately circulated statement.

The most significant criticism is that Soon and Baliunas do not present their data quantitatively—

instead they merely categorize the work of others primarily into one of two sets: either supporting or not supporting their particular definitions of a Medieval Warming Period or Little Ice Age. "I was stating outright that I'm not able to give too many quantitative details, especially in terms of aggregating all the results," Soon says.

Specifically, they define a "climatic anomaly" as a period of 50 or more years of wetness or dryness or sustained warmth (or, for the Little Ice Age, coolness). The problem is that under this broad definition a wet or dry spell would indicate a climatic anomaly even if the temperature remained perfectly constant. Soon and Baliunas are "mindful" that the Medieval Warming Period and the Little Ice Age should be defined by temperature, but "we emphasize that great bias would result if those thermal anomalies were to be dissociated" from other climatic conditions. (Asked to define "wetness" and "dryness," Soon and Baliunas say only that they "referred to the standard usage in English.")

What is more, their results were nonsynchronous: "Their analysis doesn't consider whether the warm/cold periods occurred at the same time," says Peter Stott, a climate scientist at the U.K.'s Hadley Center for Climate Prediction and Research in Bracknell. For example, if a proxy record indicated that a drier condition existed in one part of the world from 800 to 850, it would be counted as equal evidence for a Medieval Warming Period as a different proxy record that showed wetter conditions in another part of the world from 1250 to 1300. Regional conditions do not necessarily mirror the global average, Stott notes: "Iceland and Greenland had their warmest periods in the 1930s, whereas the warmest for the globe was the 1990s."

Soon and Baliunas also take issue with the IPCC by contending that the 20th century saw no unique patterns: they found few climatic anomalies in the proxy records. But they looked for 50-year-long anomalies; the last century's warming, the IPCC concludes, occurred in two periods of about 30 years each (with cooling in between). The warmest period occurred in the *late* 20th century—too short to meet Soon and Baliunas's selected requirement. The two researchers also discount thermometer readings and "give great weight to the paleo data for which the uncertainties are much greater," Stott says.

The conclusion of Soon and Baliunas that the warming during the 20th century is not unusual has engendered sharp debate and intense reactions on both sides—Soon and Baliunas responded primarily via e-mail and refused follow-up questions. The charges illustrate the polarized nature of the climate change debate in the U.S. "You'd be challenged, I'd bet, to find someone who supports the Kyoto Protocol and also thinks that this paper is good science, or someone who thinks that the paper is bad science and is opposed to Kyoto," predicts Roger Pielke, Jr., of the University of Colorado. Expect more of such flares as the stakes—and the world's temperatures—continue to rise.

David Appell is based in Lee, N.H.

Web Sites

Due to the changing nature of Internet links, the Rosen Publishing Group, Inc., has developed an online list of Web sites related to the subject of this book. This site is updated regularly. Please use this link to access the list:

http://www.rosenlinks.com/saca/wocl

For Further Reading

Alley, Richard B. *The Two-Mile Time Machine: Ice Cores, Abrupt Climate Change, and Our Future.* Princeton, NJ: Princeton University Press, 2002.

Aquado, Edward, and James E. Burt. *Understanding Weather and Climate.* 3rd. ed. Upper Saddle River, NJ: Prentice Hall, 2003.

Burroughs, William James. *Climate Change: A Multidisciplinary Approach.* New York, NY: Cambridge University Press, 2001.

Cox, John D. *Climate Crash: Abrupt Climate Change and What It Means for Our Future.* Washington, DC: Joseph Henry Press, 2005.

Fagan, Brian M. *The Little Ice Age: How Climate Made History, 1300–1850.* New York, NY: Basic Books, 2001.

Hardy, John T. *Climate Change: Causes, Effects, and Solutions.* Hoboken, NJ: John Wiley & Sons, 2003.

Lynas, Mark. *High Tide: The Truth About Our Climate Crisis.* New York, NY: Picador, 2004.

Ruddiman, William F. *Earth's Climate: Past and Future.* New York, NY: W. H. Freeman, 2000.

Speth, James Gustave. *Red Sky at Morning: America and the Crisis of the Global Environment.* New Haven, CT: Yale University Press, 2005.

Victor, David G. *Climate Change: Debating America's Policy Options*. Washington, DC: Council on Foreign Relations Press, 2004.

Weart, Spencer R. *The Discovery of Global Warming*. Cambridge, MA: Harvard University Press, 2004.

Wohlforth, Charles. *The Whale and the Supercomputer: On the Northern Front of Climate Change*. New York, NY: North Point Press, 2005.

Bibliography

Alley, Richard B. "Abrupt Climate Change." *Scientific American*, November 2004, pp. 62–69.

Alpert, Mark. "Stormy Weather." *Scientific American*, December 2004, p. 28.

Appell, David. "Behind the Hockey Stick." *Scientific American*, March 2005, pp. 34–35.

Appell, David. "Hot Words." *Scientific American*, August 2003, pp. 20–22.

Beardsley, Tim. "In the Heat of the Night." *Scientific American*, October 1998, p. 20.

Epstein, Paul R. "Is Global Warming Harmful to Health?" *Scientific American*, August 2000, pp. 50–57.

Hoffman, Paul F., and Daniel P. Schrag. "Snowball Earth." *Scientific American*, January 2000, pp. 68–75.

Holloway, Marguerite. "Core Questions." *Scientific American*, December 1993, pp. 34–36.

Karl, Thomas R, and Kevin E. Trenberth. "The Human Impact on Climate." *Scientific American*, December 1999, pp. 100–105.

Kasting, James F. "When Methane Made Climate." *Scientific American*, July 2004, pp. 78–85.

Leutwyler, Kristin. "No Global Warming?" *Scientific American*, February 1994, p. 24.

Ruddiman, William F. "How Did Humans First Alter Global Climate?" *Scientific American*, March 2005, pp. 46–53.

Simpson, Sarah. "Deserting the Sahara." *Scientific American*, October 1999, pp. 36–37.

Simpson, Sarah. "Methane Fever." *Scientific American*, February 2000, pp. 24–27.

Sturm, Matthew, Donald K. Perovich, and Mark C. Serreze. "Meltdown in the North." *Scientific American*, October 2003, pp. 60–67.

Index

A

abrupt climate change, 11,
    12–29, 31–34, 49
aerosols, 145–146
Alley, Richard B., 11
Amundsen, Roald, 119
Antarctica, 9, 15, 22, 32,
    38, 70
Appell, David, 168
Archaea, 86
Arctic, rising temperatures
    in, 108–109, 110–121

B

Baliunas, Sallie, 168, 169,
    170–173
Barnett, Tim, 171
Barton, Congressman Joe, 158
Bender, Michael L., 32
Bond, Gerard, 33, 34
Bradley, Raymond S., 159
Bralower, Timothy J., 79
Budyko, Mikhail, 55–56, 57

C

Caldeira, Kenneth, 59
Cambrian explosion, 53, 67
carbon dioxide, 7, 30, 33, 36,
    37, 38, 40–41, 45, 48,
51, 57, 58, 59, 60–63,
    64, 65, 70–71, 78, 81,
    111, 116, 117, 142, 146,
    150, 153, 160
   and deforestation, 43–44
   1991 drop in levels of, 164,
    165–168
   and prehistoric warming,
    81–82, 84–88, 90, 91,
    93–94, 97, 99
   and tropical forests,
    100–103
cholera, 137, 138, 140, 141
Clark, David B., 100
Clark, Deborah A., 100,
    102, 103
Claussen, Martin, 74–75,
    76, 77
climate change, abrupt, 11,
    12–29, 31–34, 49
climate modeling, 9, 33, 45,
    143, 144, 147–155
conveyor belt circulation,
    explanation of, 21
cyanobacteria, 57, 82

D

Day After Tomorrow, The, 12
deforestation, CO_2 and, 43–44

dengue fever, 122, 126, 128, 129, 130, 140

Dickens, Gerald, 78, 79, 80, 81

Doomsday Book, 44

dust bowl, U.S., 16

E

Earth Observing System, 152

Ecological Society of America, 100

Eemian, 32

Eisner, James B., 107, 108

El Niño, 16, 20, 106, 107, 134–136, 158, 163, 165, 167

Emanuel, Kerry A., 104, 107

encephalitis, 126, 132

Epstein, Paul, 121–122

eukaryotes, 67–69

Europa, methanogens on, 91

F

farming/agriculture, and effect on climate, 42–47

foraminifera/forams, 77–78, 79–80

fossil fuels, 57, 78, 100, 141, 145, 146, 147, 163, 166, 168

G

Global Business Network, 25–26

global warming
and human health, 122–143

and hurricanes, 9–10, 104–105, 105–108

Gray, William M., 106, 107

greenhouse gases, 7, 8, 11, 34, 35, 36, 37, 38, 41, 45–46, 47, 48, 51, 64, 78, 82, 84, 85, 89–90, 94, 100, 116, 142, 144, 146, 153, 155, 163, 164, 168, 170

greenhouse warming, basic mechanics of, 7

Greenland, 13–15, 16, 18–21, 22, 29, 30, 31–34, 38, 70, 110, 114, 144, 173

Greenland Ice-Core Project (GRIP), 32, 33

Greenland Ice Sheet Project II (GISP2), 32–33

H

Halverson, Galen Pippa, 65

hantavirus pulmonary syndrome, 133–136

Harland, W. Brian, 54–55, 67

hematite, 95

Hoffman, Paul F., 49

Hughes, Malcolm K., 159, 171

human health, global warming and, 122–143

Huronian glaciation, 95, 96

hurricanes
Charley, 9, 104
Frances, 9, 104
Ivan, 9, 104
Jeanne, 9, 104

Katrina, 10, 104
Mitch, 138
Rita, 10, 104
hurricanes, global warming
and, 9–10, 104–105,
105–108

I

ice ages, 12, 14, 16, 25, 36,
38, 46–47, 48, 50–51,
55, 59, 69, 71, 78, 82,
84, 95, 147
ice-albedo feedback, 56,
111, 118
Inhofe, Senator James M., 161
Intergovernmental Panel on
Climate Change, 24, 142,
144–145, 160, 169, 173

J

Jones, Philip D., 159, 171

K

Karl, Thomas R., 143, 144
Kasting, James F., 59, 81–82
Katz, Miriam E., 78, 79,
80, 81
Kaufman, Alan Jay, 65
Keeling, Charles D., 100,
101, 165, 166
Keeling, Ralph F., 166,
167, 168
Kininmonth, William, 170
Kirschvink, Joseph L.,
58–59, 65
Kubatzki, Claudia, 76

Kutzbach, John E., 45, 75, 76
Kyoto Protocol/Treaty, 155,
163, 169, 173

L

La Niña, 106, 134, 135,
136, 163
La Selva Biological
Station, 100
Lehman, Scott J., 33, 34
Little Ice Age, 15, 25, 29,
112, 119, 120–121, 161,
170, 172

M

malaria, 122, 126, 127,
128–129, 138, 140
Mann, Michael, 157, 158,
158–164, 171
Mars, methanogens on, 91
Mayan civilization, end of, 16
Medieval Warming Period,
161, 170, 172
meta-analysis, explanation
of, 169
methane, 7, 14, 36, 37, 38,
39, 40, 41, 42, 77–78,
78–81, 116
and farming, 42–43, 45
and prehistoric warming,
82–98
methanogens, 82, 83, 86,
87–95, 96, 97
on Europa, 91
on Mars, 91
Morgan, M. Granger, 31

mosquitoes, 122, 126–128, 128–130, 131, 132–133, 135, 138, 139, 140

Mullen, George H., 84

N

National Academy of Sciences, 158

National Aeronautics and Space Administration (NASA), 87, 94, 97, 152, 160

National Oceanic and Atmospheric Administration, 152, 166

negative feedback cycles, explanation of, 7–8, 111

Neoproterozoic period, 51, 52–53, 54–55, 58, 59, 66, 67, 69, 70

Norris, Richard D., 80

North Atlantic conveyor, 21–23, 24, 25, 26

North Atlantic Oscillation, 107

O

Oppenheimer, Michael, 171

P

Pak, Dorothy K., 79

Paleocene period, 78, 80, 81

permafrost, 114–115

Perovich, Donald K., 117–118

Pielke, Roger, Jr., 173

Pierrehumbert, Raymond T., 64

Pinatubo, Mount, 150, 164, 166, 167

Piper, Stephen C., 100

positive feedback cycles, explanation of, 8, 111

precession, 39, 40

prehistoric warming
  carbon dioxide and, 81–82, 84–87, 88, 90, 93–94, 97, 99
  methane and, 82–98

Proterozoic period, 95, 96

R

Randerson, James T., 102

Rift Valley fever, 138, 141

rising sea levels, 10, 113–114, 118, 123, 154, 164

Ruddiman, William F., 35

Rudwick, Martin J. S., 67, 69

Rye, Rob, 85

S

Sagan, Carl, 84

Sahara, computer simulations of creation of, 73, 74–77

schistosomiasis, 125

Schmidt, Gavin A., 160–161, 162

Schrag, Daniel P., 49

SEARCH (Study of Environmental Arctic Change), 120

Serreze, Mark C., 108–109, 113, 118

siderite, 85

Siefert, Janet L., 87, 90

snowball earth hypothesis, 50–71

Soon, Willie, 168, 169, 170–173

Stott, Peter, 172–173

Sturm, Matthew, 109–110, 112

Suess, Erwin, 81

T

Tans, Pieter P., 166, 168

Titan, 83, 94

Trenberth, Kevin E., 143, 144

tropical trees, effect of warmer nighttime temperatures on, 99, 100–103

U

United Nations Environmental Program, 145

United Nations Framework Convention on Climate Change, 155

U.S. Department of Defense, 25

U.S. National Research Council, 28

V

Vavrus, Stephen J., 45

Vikings, 15, 29

Vostok, 32, 38, 39

W

Weaver, Andrew J., 34

West Nile virus, 124, 132, 139, 140

Wexler, Harry, 165

World Meteorological Organization, 144–145

Y

yellow fever, 126

About the Editor

Katy Human writes about science for the *Denver Post*, covering everything from space science to wildlife research. Human has a Ph.D. in ecology from Stanford University and has been fascinated by ecology and climatology since her childhood, when she spent many days in the Smithsonian's Natural History Museum in Washington, D.C. She has been a journalist since 1997.

Illustration Credits

Cover: Slim Allagui/AFP/Getty Images; p. 17 (top graph) Eliza Jewett (Sources: M. Stuiver et al., and K. Cuffey et al.); p. 17 (bottom graph) D. Hodell et al. (Source: National Geophysical Data Center); pp. 22, 23, 60, 61, 62, 63 David Fierstein; pp. 54, 68 Heidi Noland; pp. 86, 92 Johnny Johnson; pp. 89, 90, 91 Don Dixon; p. 127 Bryan Christie (Source: Pim Martens, Maastricht University); pp. 131, 139 Bryan Christie; p. 134 Bryan Christie (Sources: NOAA Climate Prediction Center (http://chge2.med.harvard.edu/enso/disease.html); p. 162 (Source: Intergovernmental Panel on Climate Change).

Series Designer: Tahara Anderson